ROBERT FLUDD

ESSENTIAL READINGS

William H. Huffman, Ph.D.,
is a native of St Louis, Missouri, and
presently resides in Charlotte, North
Carolina. His lifelong interest in the
esoteric eventually led him to doctoral
studies in the history of philosophy,
particularly the Platonic, and a
concentration on the Renaissance
Hermetic tradition. Under the
inspiration of Frances Yates and others,
he completed a study, *Robert Fludd and
the End of the Renaissance*, which was
published in 1988. He believes that
Fludd's books represent a grand
summation of centuries of Christian
Neoplatonist Hermeticism and will
remain, with their unparalleled
illustrations, a feat perhaps never
accomplished again.

ROBERT FLUDD

ESSENTIAL READINGS

Selected and Edited by
WILLIAM H. HUFFMAN

Aquarian/Thorsons
An Imprint of HarperCollins*Publishers*

The Aquarian Press
An Imprint of HarperCollins*Publishers*
77—85 Fulham Palace Road,
Hammersmith, London W6 8JB

Published by The Aquarian Press 1992
1 3 5 7 9 10 8 6 4 2

Selection and Introduction © William H. Huffman 1992

A catalogue record for this book
is available from the British Library

ISBN 1 85538 142 7

Typeset by Harper Phototypesetters Limited,
Northampton, England
Printed in Great Britain by
Billings & Sons Ltd, Worcester

Essential Readings

This series is designed as an introduction to the life and works of major figures in the history of ideas, particularly in the realm of metaphysics and the esoteric tradition. This anthology of Robert Fludd's writings represents an important addition to the series, particularly those volumes devoted to sixteenth and seventeenth-century philosophers and mystics, including John Dee, Paracelsus and Boehme.

Robert Fludd was one of the last 'Renaissance men', who sought to embrace the whole of human knowledge within a divine and hierarchically ordered cosmology. Born in Elizabethan England and educated at Oxford University, he travelled subsequently on the Continent where he came into contact with occult and alchemical ideas among the supporters of Paracelsus (1493-1541), the Neoplatonist and medical reformer. Establishing himself in London as a physician, partly on Paracelsian principles, Fludd made his own philosophical debut with three small books in defence of the contemporary Rosicrucian manifestos (1614-16) and their doctrines of macro-microcosmic harmony, magic, the Cabala, and use of the Hermetic texts.

Fludd's early identification with Rosicrucianism indicates his fundamental allegiance to that complex of Neoplatonic,

Hermetic and cabalistic ideas which had profoundly influenced the Renaissance mind from the mid-fifteenth to the early seventeenth century. In his later masterpiece *The History of the Macrocosm and Microcosm* (1617-26), Fludd elaborated the major and final statement of this Rosicrucian-Neoplatonic-Hermetic philosophy, which combined a practical examination of nature with a spiritual view of the universe as an intelligent hierarchy of beings in three orders (empyrean or angelic; ethereal or stellar; elemental or material). This eclectic philosophy recognized the universality of truth and drew its wisdom from all possible sources — Catholic and Protestant, the Hebrew Bible, Plato, Pythagoras and Hermes Trismegistus. Above all, this philosophy provided for human enquiry and science within a meaningful and divine Creation; its proper end was the direct knowledge of God.

Professor Huffman provides the reader with a concise introduction to Fludd's life and an accessible selection of his works, including the famous *History of the Macrocosm and Microcosm*, his declaration to King James I, his patron, controversial and polemical treatises, and later books. These are illustrated by a number of beautiful engravings, which Fludd used throughout his works to such great advantage in symbolizing and condensing the complex and fascinating hierarchies of his cosmology with reference to music, astronomy, meteorology, mechanics and human anatomy. This anthology of the 'greatest summarizer and synthesizer' of the Renaissance Neoplatonic tradition offers us a valuable glimpse of scientific thought poised between the medieval and modern world-views, still focused on the unity and order of all things rather than on an endless inventory of their properties.

NICHOLAS GOODRICK-CLARKE
April, 1991

Contents

List of Illustrations

Acknowledgements

I would like to gratefully acknowledge the kind permission of the following:

Ambix, The Journal of the Society for the Study of Alchemy and Chemistry, to reprint excerpts from 'Truth's Golden Harrow: An unpublished Alchemical Treatise by Robert Fludd', edited with an introduction by C. H. Josten, Vol. 3, Nos 3 and 4 (April 1949), 91-150; and 'Robert Fludd's "Declaratio Brevis" to James I', translated by Robert A. Seelinger, Jr, edited with an introduction by William H. Huffman, Vol. 25, Part II (July 1978), 69-92.

Excerpts from Pauli, W., 'The Influence of Archetypal Ideas on the Scientific Theories of Kepler', translated by Priscilla Silz, in *The Interpretation of Nature and the Psyche* (Pantheon Books, New York, 1955), Bollingen Series LI, Copyright 1955 © renewed by Princeton University Press, pages 190-236, reprinted by permission of Princeton University Press.

Excerpts from Fludd's *Utriusque cosmi historia* in *The Origin and Structure of the Cosmos*, translated by Patricia Tahil, edited with an introduction by Adam McLean, Magnum Opus Hermetic Sourceworks Number 13, reprinted by permission of Phanes Press.

The remaining excerpts are from my own collection.

Introduction

Robert Fludd was the most prominent Renaissance Christian Neoplatonist alchemist of his time, and the greatest summarizer and synthesizer of that tradition of his age. There were several major influences in his life: 1) his profoundly held spiritual beliefs; 2) his family heritage; 3) his broad Renaissance education, which centred on the Platonic-Neoplatonic philosophical outlook allied with alchemy and Paracelsian medicine; 4) his steadfast belief in the comprehensible, integrated nature of the universe and everything in it. The first was expressed through his insistence on the true wisdom of God being the basis for his philosophy and his frequent reference to the Bible as proof of physical and metaphysical principles; the second through the proud display of his coat of arms in his two portraits in his books, his church monument, and his remarks in *Dr Fludds Answer unto M. Foster*; the third through the astonishing breadth of his knowledge, which even led to the accusations of his having help from the Rosicrucians in writing his *History* (see 'A Philosophical Key', Chapter 4); and the last through the monumental attempt in his works to show the exact details of how the universe and everything in it came into being, related to one another, and carried out its assigned role; all of which was in accord with esoteric sense of the Bible, the Cabala, musical harmonies and alchemical axioms. With Fludd, the Renaissance Christian Neoplatonist/Hermetic tradition that began with Marsilio Ficino in Florence reached its greatest expression. His legacy of voluminous written works, with unsurpassed illustrations, and his engagement with some of the leading figures of the day make him stand out as a key to understanding the world of the early seventeenth century.

After Fludd's death, the forces already at work bringing about a sea change in Western thought, commonly called the Scientific Revolution, became prevalent, and both he and his world faded into the background. Until the 1960s — except for the pioneering post-war work of C. H. Josten — Fludd was usually considered just an occultist Rosicrucian, and was only of interest to those concerned with esoteric subjects, such as Mme Blavatsky, the Theosophical Society founder, and the writer Arthur Edward Waite (1857-1942). In the mid-sixties, that changed, due to two pioneering books: Frances Yates' *Giordano Bruno and the Hermetic Tradition* (1964) and Allen Debus' *The English Paracelsians* (1965) (see Bibliography for full details). The former showed the influence of the writings attributed to Hermes and, indeed, the influence of the entire Neoplatonic renaissance that stemmed from the work of Marsilio Ficino in late fifteenth-century Florence. Its impact on Fludd is clear. Dame Frances continued to add intriguing fact and speculation to the field, and created a new perspective on the era from the history of thought. Allen Debus' approach from the history of science and medicine also did much to restore Fludd as an important figure of the early Scientific Revolution. His several books and many articles will no doubt assure continued recognition of Fludd's significance for his time. My own book, *Robert Fludd and the End of the Renaissance* (Routledge, 1988), owes much to the pioneering work of both.

Background

Fludd was born at the family's country manor in Kent, Milgate House, a few days before 17 January 1573/4, when he was baptized at the Bearsted Parish Church. Descended from gentry on both sides of his family, he proudly displayed his coat of arms in his portraits and placed the title of Esquire before Doctor as being most appropriate for one of gentle birth.

The Fludds were originally from Wales, but had made their home in Shropshire for some three generations or more before Robert's father, Sir Thomas Fludd, acquired Milgate House in the mid-1560s and married Elizabeth, the daughter of Philip Andrews, Esquire, of Wellington in Somerset.[1] Sir Thomas built his fortunes on service to Queen Elizabeth I, beginning as a victualler to Her Majesty's troops in the garrison towns of Berwick and Newhaven in the early 1560s. After he acquired Milgate House, he was appointed Justice of the Peace for the Bearsted district, a post he held until his death in 1607. In 1568 the Queen also granted him lifelong tenure of the post of Surveyor of Crown Lands in Kent and the cities of Rochester and Canterbury, and in the mid-1580s made him Receiver of Crown Revenues for Kent, Surrey and Sussex. In 1589, Elizabeth sent 4,000 troops to France to aid the French King and appointed Sir Thomas paymaster. It was this latter service that won him knighthood in January 1589/90 when the English force returned. On the accession of James I in 1603, his stewardship of some 20 manors in Kent was renewed by letters patent. Sir Thomas also represented the borough of Maidstone, Kent, in three parliaments, 1583, 1597–8 and 1601. Trained in the law, he was given the honour of admission to Gray's Inn on 9 March 1600/01.[2]

Sir Thomas and Elizabeth Fludd had twelve children in all, nine of whom lived to adulthood. There were six sons: Edward, Thomas, William, John, Robert and Philip; and three daughters: Joan, Katherine and Sarah. Elizabeth Fludd died in 1592, the year Robert entered Oxford, and Sir Thomas was remarried to Barbara, daughter of Matthew Bradbury, Esquire, of Essex, the widow of Sir Henry Cutts. When Sir Thomas died in 1607, only Thomas, John, Robert and the three daughters were still living.[3]

Robert was the seventh child, and very likely received his

early education from tutors at home. It may have been some now unknown tutor who intrigued the unusually studious youngster with the mysteries of the universe and started him on his personal journey of discovery. On 10 November 1532, at the age of 18, Robert followed his older brother John in matriculating at St John's College, Oxford. But unlike his older brothers, he decided not to study the law, and was enrolled in the Arts faculty. His tutor was John Perrin (1558-1615), who became a Doctor of Divinity and was the lecturer in Greek at the college.[4]

At that time, it was unusual for sons of knights and peers to actually graduate, even though it was fashionable to attend university for a while; but Fludd was clearly of a serious scholarly bent, and finished his BA in 1596, and MA in 1598. It was at St John's that he further developed his interest in the Neoplatonic view and began systematizing his studies, works that would later appear as chapters in his books. Fludd relates a story of his tutor, Perrin, bursting into his room while he was working on his treatise on music, and demanding that he use his astrology to find out who had robbed him; he obliged.

Although Aristotelian structure and content was strongly present in the university education of the day, there was a significant infiltration of Platonism, and from the story of the tutor, the occult arts were clearly used by Fludd and were known to his tutor. It was also from this period that Fludd's interest in medicine derived, because St John's was one of the few colleges at Oxford that had a medical Fellow in residence, and the President of the College while Fludd was there (Ralf Hutchinson) had in fact been the medical Fellow; he later became a theologian. Because of its strong theological content and orientation, Fludd's undergraduate education also strongly reinforced the framework of his Judaeo-Christian heritage.[5]

Travels Abroad

It was accepted that part of a well-bred gentleman's education should include foreign travel, so, in July 1598, when Fludd received his MA, he wrote in the register that he was going abroad. For most of the next six years, he travelled in continental Europe, tutoring, writing and visiting places of interest to him, including France, Spain, Italy and Germany. From different passages in his books, a likely itinerary would have taken him from England to Paris for a time, where he met Lord Bourdalone, the Chief Secretary to Charles of Lorraine, fourth Duke of Guise; Bourdalone and Fludd became good friends. From there he stopped at Lyons, then headed south, but his passage to Italy was blocked by snow in the St Bernard Pass, so he passed the winter of 1601-2 in Avignon. Here he wrote his treatise on geomancy for the Papal Vice-Legate, and those on the art of motion and astrology for Reginaud of Avignon. He also dedicated tracts on music and the art of memory to another noble student, the Marquis de Orizon, Viscomte de Cadenet. When winter passed, Fludd was summoned to Marseilles, where he instructed the Duke of Guise and his brother Francois, a Knight of Malta, in mathematics. He wrote the sections of his *Macrocosm* history on arithmetic, geometry, perspective and the 'secret military arts' for Francois' instruction. Sometime after that, he went on to Provence, then crossed over into Italy, where he travelled through Livorno to Rome. There he said he met a Master Gruter, a Swiss, who taught him a great deal. Then he appears to have gone to Venice, and on to Augsburg. While in Germany, Fludd would no doubt have visited the court of the Elector Palatine in Heidelberg, which had a famous library and warm relations with England; and the Court of Moritz the Learned, Landgrave of Hesse, in Kassel. At some point, he travelled in Spain, but exactly when is not known.[6]

Medical Practice

In late 1604 or early 1605, Fludd entered Christ Church College, Oxford, to take degrees in medicine, and on 16 May 1605, he was granted a MB, MD and licence to practise. After graduation, he set up a medical practice in London and applied for admission to the College of Physicians, the governing body of medical affairs in the city. After some heated exchanges with some of the Censors of the College about Paracelsian vs Galenical medicine and several examinations, he was admitted as a Fellow on 20 September 1609. He himself served as Censor in 1618, 1627, 1633 and 1634, and enjoyed the company of his good friends William Harvey, Richard Andrewes (who was a classmate of his at St John's, Oxford) and Sir William Paddy, to whom Fludd dedicated his medical work, the *Medicina Catholica* (Catholic Medicine) of 1629. Fludd gave the public anatomy lecture at the college at least once, in 1620, and actively participated in its work. He showed his pride in being a member by donating a set of his works for the library in 1635, leaving it £20 in his will, and being the first Fellow known to mention his membership on his church monument. [7]

Fludd practised medicine from his house in Fenchurch Street, and had a resident apothecary. From various threads of evidence, it would appear that he was well versed in Paracelsian medicine and its attendant Christian Neoplatonic-Hermetic–alchemical metaphysics, but in fact employed much standard Galenical medicine of the day in his practice. In 1618, the College of Physicians issued its book of standard pharmacology to be followed in the city, the *Pharmacopoeia Londinensis*, and Fludd is listed at the beginning with the other members as an author. Since this was one of the areas governed by the College, Fludd's practice would clearly follow this standard. [8]

Fludd's theory of disease was that a wind controlled by an evil spirit excites lesser evil spirits in the air that are always present. The latter may enter the body through the pores or from breathing; and, if there is a weakness somewhere in the body, the evil forces will attack the bonds that hold the four humours in balance, which causes a malady. If it is not properly attended to, the imbalance will become so violent that it will cause the entire physical body to collapse. Treatment consists of restoring the imbalance by providing, through a sympathetic compound, the angelic forces that were weak and allowed the malady to occur in the first place. The administered medicine worked through sympathetic natural magic, and could consist of appropriate herbs gathered under the right astrological influences, chemical medicines or 'magnetic' preparations, such as the weapon-salve, that worked at a distance.[9]

Alchemical Laboratory

In 1606, Fludd returned to the Continent, which we know from a letter written by Sir Thomas Fludd to the Earl of Salisbury:

> 1606, August 20 — A Son of mine being a doctor of physic and greatly desirous to have conference with certain physicians, Italian and French, now in France, his acquaintance and good friends in his travel beyond seas, touching secrets and other things concerning that study, and to return within two months, about a month past went over into France and landed at Dieppe, from whence he wrote the letter herein . . .

Sir Thomas included the only known letter of Robert Fludd's in order to alert the Earl about a possible place of entry by traitors into England. Fludd had been detained briefly on the English coast because of his resemblance to a suspect, and he found out more from boatmen who were lodgers at the same inn, the 'Fleur de Lewse' in Dieppe. It also contains a description of Fludd:

> Robert Fludd to Sir Thomas Fludd, his father.
> I was examined lying at Hide [Hythe] because I was of the description of him that was searched after, for he had, they said, a small stature, lean visage, auburn hair, etc . . .[10]

It may have been on this trip or a subsequent one that Fludd recruited a French 'operateur' or technician to set up and run an alchemical laboratory for him in London. The main proof of his view of the origin and dynamics of the universe, he maintained, was his alchemical experiments with wheat, and since we know that his *History of the Microcosm and Macrocosm* was finished about 1610-11, he must therefore have set up his laboratory between about 1606 and 1610. [11]

Associations

Fludd was associated with three circles in England and various friends on the Continent. The first of his circles encompassed his medical colleagues in the College of Physicians in London. Among the most notable of these were William Harvey and Sir William Paddy. Fludd witnessed a number of Harvey's anatomical investigations, and was the first to agree with his theory of the circulation of the blood in print. It has even been suggested that Harvey may have derived his theory of blood circulation from Fludd's metaphysics. [12] Sir William Paddy was James I's Physician-in-Ordinary, to whom Fludd dedicated the first part of his *Medicina Catholica*. Fludd also tells us that another medical colleague, Richard Andrewes, was his 'worthy friend,' and had read his *History of the Macrocosm* manuscript about 1610.

The second circle included those with alchemical interests, the most prominent of whom was John Thornborough, the Bishop of Worcester at Hartlebury Castle, which was described by a contemporary as 'an Apollinian Retreat, as a living library, a flourishing Academy, or a religious abbey'. [13] Fludd was a good friend of the Bishop and his son, Sir Thomas Thornborough, and was a guest at the castle on at least one occasion. Fludd dedicated his *Anatomiae Amphitheatrum* to the Bishop, and the latter's treatise on the Philosopher's Stone of 1621 contains four references to Fludd's *Tractatus Theologo-Philosophicus*, which included praise for the Rosicrucians. Fludd dedicated the *Philosophia Sacra* of 1626 to another eminent cleric, John Williams, Bishop of Lincoln.

The third circle of Fludd's associates centred around Sir Robert Bruce Cotton (1571-1623) and his renowned library, which, among many other works, included the remains of John Dee's library. Fludd dedicated the second part of his *Medicina Catholica* to Cotton, whom he calls his 'singular friend'. This circle included the scholars and antiquaries William Camden (1551-1623) and John Selden (1584-1654); the latter praised Fludd's medical skill in curing him of a sickness in the dedication of the first edition of his *Titles of Honour* of 1614.

On the Continent, we know that Fludd met many people on his six-year sojourn and stayed in contact with them, as seen in Sir Thomas Fludd's letter above and those attached to the 'Declaratio Brevis' (see below and Chapter 3).[14]

Written Works and Controversies

Fludd tells us that he was at work on his treatise on music at Oxford in the 1590s and was already well versed in astrology. To tutor Charles of Lorraine and his brother Francois, he wrote treatises on arithmetic, geometry, perspective and 'the secret military arts' in about 1601-2. Through the early 1600s, both during and after his continental travels, he continued to write, and completed his *History of the Macrocosm* by about 1610-11, after which he no doubt went to work on the *Microcosm*. By 1614, Fludd had reached the age of 40, was well established in his London medical practice and alchemical experiments, and had over 20 years of his philosophy written in manuscript.[15] But the event that propelled him into print was the furor over publication of the Rosicrucian manifestos beginning in that year. In 1614, from the press of Wilhelm Wessel in Kassel came a book with the title:

> Universal and General Reformation of the Whole Wide World; together with the *Fama Fraternitatis* of the Laudable Order of the Rosy Cross, written to all the Learned and Rulers of Europe; also a short Reply sent by Herr Haselmeyer, for which he was seized by the Jesuits and put in irons on a Galley; Now put forth in print and communicated to all true Hearts.[16]

The 'General Reformation' is a satire by Trajano Boccalini of 1612, which doubts that a general reformation can ever be put into place by relying on the learned establishment. The *Fama* tells a marvellous mythological tale of Brother R.C., an initiate into the divine mysteries, who founded the secret Fraternity of the Rosy Cross after failing to convince the learned establishment of the truth of his knowledge; and expresses the desire to make the Fraternity's presence known now because of the rediscovery of Brother R.C.'s wondrous tomb. The Haselmeyer letter says that he saw the *Fama* in manuscript in the Tyrol in 1610. In early 1615, the *Fama* was followed by the *Confessio Fraternitatis R.C., Ad Eruditos Europae* from the same publisher, a work which continued the story, but in more didactic language. Later in 1615 Wessel published the *Fama* and *Confessio* together in one volume. In the *Fama*, those of like mind and heart are invited to join the Fraternity:

> So according the will and meaning of *Fra: C.R.C.* we his brethren request again all the learned in Europe, who shall read (sent forth in five Languages) this our *Famam* and *Confessionem*, that it would please them with good deliberation to ponder this our offer, and to examine most nearly and most sharply their Arts, and behold the present time with all diligence, and to declare their mind, either *Communicato consilio*, or *singulatim* by Print.
>
> And although at this time we make no mention either of our names, or meetings, yet nevertheless every one's opinion shall assuredly come to our hands, in what language so ever it be; nor any body shall fail, who so gives but his name to speak with some of us, either by word of mouth, or else if there be some let in writing. [17]

Just about everything in the manifestos would appeal to Fludd: their purpose of reforming learning, medicine, religion and politics, of revealing all the divine secrets of nature to the worthy, and of restoring union and concord in Europe along Protestant Hermetic-Paracelsian lines. They were, in effect, an expression of the Renaissance ideal.

The European presses were kept busy for the next few years publishing replies to the manifestos, both for and against, as well as supplications for membership in the august order. One serious reply against the Brotherhood that caught

Fludd's attention came from Andreas Libavius (1540-1616), a German Lutheran chemist. Libavius had established his reputation by publishing a plain-language textbook on chemistry and chemical medicines, and engaged in the chemical debates of the day. In several works, he attacked the Fraternity, primarily for heresy, sedition and use of diabolical magic to effect their results. Fludd saw Libavius' *Analysis of the Confession of the Fraternity of the Rosy Cross* (Frankfurt, 1615), and sketched a quick reply that could be sold at Frankfurt book fair in the spring of 1616, the *Apologia Compendiaria, or a Brief Apology, washing away and cleansing the stain of suspicion and infamy applied to the Fraternity of the Rosy Cross with, as it were, a Fludd of truth* (see Chapter 1). The *Apology* outlines a longer work to appear the following year, and contains an epilogue to the Brothers R.C., humbly requesting consideration as a candidate for their order. Fludd would in fact have been ideal for membership.[18]

In 1617, Fludd published three books: the promised longer apology for the Rosicrucians, a small mystical work on life, death and resurrection, and the first volume of his *magnum opus*. The apology for the Rosicrucians, the *Tractatus apologeticus integritatem Societatis de Rosea Cruce defendens*, included the previous year's *Apology* as a proemium. It rebuts Libavius' *Analysis* by juxtaposing passages from it with ones from the *Confessio*, and asserts that magic, the Cabala and astrology are true arts if rightly done. Fludd deplores the state of teaching of natural philosophy, medicine, alchemy, mathematics and the moral arts in the universities, which rely too heavily on the superficiality of Aristotle and Galen while the true essence of things is ignored, and asserts that the key to understanding how all things are related comes from an occult knowledge of the musical harmony throughout the universe. Last of all, he shows how the 'divine light' is the central animating force in the created universe, and how many wondrous effects may be manifested at a distance through divine principle and without diabolical help. The *Tractatus apologeticus* closes with the same letter to the Rosicrucians with this addition:

Farewell, Brothers most dear, in the name of those whom you sincerely honour. Farewell, I say, and farewell again; favour me and approach me (I implore and entreat you by your assurance and because of the ignorance of the age in the true and pure Philosophy); be mindful of me and your promises.

Both the apologies were published in Leiden by Godfrey Basson, the son of an Englishman and a lover of occult subjects.[19]

Fludd's other two books of 1617 came from the press of Johann-Theodor de Bry, with whom Fludd enjoyed a long and fruitful relationship. The first was a small work entitled *Tractatus Theologo-Philosophicus . . . de Vita Morte et Resurrectione*, which he penned under the anagrammatic pseudonym 'Rudolfo Otreb, Britanno'. This work may have been in progress when the Rosicrucian manifestos appeared. The small book was dedicated to the Brothers R.C., and they were mentioned in 10 of the 126 pages, which deal with life, death and resurrection along the characteristic lines of Fludd's philosophy. In the section on life, Fludd maintains that the arcane mysteries of the patriarchs, prophets and apostles remain among men, and that there is a terrestrial paradise as well as celestial. The key to the hidden wisdom of Moses and Elias is still present among certain elect of clean and pure heart, who have superhuman powers, but remain unknown to the ordinary world and live in their secret houses of the Holy Spirit.[20]

Having said this, Fludd addresses 'the most illuminated Brothers', saying he has diligently scrutinized the *Fama* and *Confessio* with his eyes and his soul, and concluded that they are in fact illuminated by the divine spirit, and that the Brothers have been shown the things which were prophesied mystically in the Scriptures as preceding the end of the world. If their actions are in consonance with their words, then their predictions must be believed, since they conform to sacred truth. The Brothers are praised for being able to make gold, but yet disdaining the earthly kind for the heavenly. Fludd issues an invitation to all humanity, whose minds are befogged with ignorance, to join with him in acknowledging that the Fraternity of the Rosy Cross is, without doubt, under the guidance of the sacred spirit, and that the Brothers reside at the summit of a high mountain where they may inhale the most sweet and rarefied aura of the psyche, which is the adornment of true wisdom.

In the section on resurrection, Fludd compares the gold of the impious and spurious alchemists to the heavenly kind spoken of by the Brothers. Indeed, the great mystery of the resurrection is shown by the most illuminated Fraternity in the description of Brother R.C.'s tomb in the *Fama*: the incorruptible wisdom of Christ is represented by the sun in the ceiling of the tomb, which shines down on his undecayed body. [21]

Given Fludd's two Rosicrucian apologies in his own name and the *Tractatus Theologo-Philosophicus*, his probable links with Michael Maier, the other major Rosicrucian apologist (see below) and the similarity of his thought and that of the manifestos, it is not surprising that he would have been considered an initiate of the Fraternity. But Fludd's own words on the matter, given as an oath in the unpublished 'A Philosophical Key' (see Chapter 4) clearly lay that centuries-old idea to rest. [22]

The third book of 1617 was far and away the most important: the first volume of the *History of the Microcosm and Macrocosm*, which was dedicated to God and James I and dealt with the creation, structure and dynamics of the Macrocosm (see Chapter 2 and Bibliography). It was followed the next year by the second part, the *De Naturae Simia*, which covered the arts and sciences of man, through which he becomes the 'ape of nature'.

Thus, by 1618, Fludd had published two overt and one covert apologies for the Rosicrucians and the two volumes of his *History of the Macrocosm*; the ramifications were to prove interesting. That same year he also was elected a Censor of the College of Physicians, and on 12 May, two holders of the monopoly patent for making steel in England complained to the Privy Council that '. . . one Robert Floude, doctor of physicke, hath entertayned a Frenchman to practise the same, hath sett up furnaces, and is now in worke [making steel]'. The Council directed the Barons of the Exchequer to suppress the unwarranted steel-making as it saw fit. [23] Fludd's reply would come two years later.

Apparently later in 1618 or early 1619, 'some envious persons' called to James I's attention that Fludd had written a work dedicated to him containing spurious philosophy, and that he was an unchaste man who had also defended the Rosicrucians, who were guilty of heresy and religious innovation. The King thereupon called Fludd before him to explain his works. The audience apparently went very well, as Fludd reported later in *Dr Fludds answer unto M. Foster* (see Chapter 7), and James became Fludd's patron. In the course of the interview, James must have suggested that Fludd write out a digest of his defence, which he immediately set out to do in late 1618 or early 1619, the result being his 'Brief Declaration to James I' (see Chapter 3).

In this Fludd denies that the Rosicrucians are guilty of religious innovation and heresy because they are acknowledged by German Catholics and Lutherans to be Calvinists, and he himself has always been a steadfast adherent to the reformed religion of England. As to the charge he was unchaste, he swears before God and His Majesty that he is a *virgo imaculata*. He defended the Rosicrucians because they advocated 'the revelation of the true basis of natural philosophy, commonly unknown to this day, and the discovery of the profound secret of medicine . . . ' To add credibility to his defence of the *Macrocosm* history, Fludd includes letters of praise from the Continent, the most important of which is from Dr Gregor Horst (see notes to Chapter 3).

Another piece of intriguing information comes from the 'Declaratio': Fludd explains that the dedication of the *Macrocosm* to God and James I came after some controversy 'between the individual to whom I entrusted this volume in England and the engraver and printer concerning the dedication of my work'. It seems that the courier wanted to dedicate the book to the Landgrave of Hesse, while the printer and engraver (Johann-Theodor de Bry and Hieronymus Gallerus) wanted to give it to the Count Palatine, so Fludd stepped in and settled the matter. It would appear that the 'trusted individual' was Michael Maier (*c.*1568-1622). Maier made a trip to England after 1612, and made a number of acquaintances, including Sir William Paddy (to whom he dedicated a book, as did Fludd) and very likely Fludd himself. In his *Atalanta Fugiens* of 1618, Maier mentioned a *Tractatus de Tritico* which is probably Fludd's treatise on his experiment with wheat (see 'A Philosophical Key', Chapter 4).[24] Since Maier was the other main Rosicrucian apologist (*Silentium Post Clamores*, 1617; *Symbola Aureae Mensae*, 1617; *Themis Aurea*, 1618), was also published by de Bry and dedicated three of his own works to the Landgrave of Hesse, whose service he was in for a time, it is quite likely it was he who carried Fludd's manuscripts to de Bry and started their long relationship.

Whether James I read the 'Declaratio' we do not know, but it is in the Royal Manuscripts collection of the British Library and provides us with an interesting look at Fludd's views as presented to the King. We do know that it was during this time of controversy that James became Fludd's patron and had the Privy Council consider his application for a patent (see below).[25]

In 1619, the first section of Tract One of the *Microcosm* history (dedicated to James I) appeared from de Bry in Oppenheim, and Fludd went to work writing 'A Philosophical Key' as a sequel to the 'Brief Declaration' (see Chapter 4). No doubt he felt that a response should be made to those who had denounced his work to James in the first place and others of their kind. The new work would also give him the opportunity to explain his metaphysics in a more concise way and show its underpinnings with his experiments on wheat. It is interesting that in this manuscript, kept at Trinity College, Cambridge, unknown until recent years and published only in 1979, Fludd expresses chagrin about his Rosicrucian apologies: they did nothing to attract the attention of the Brothers R.C. (the same experience that all the applicants had);

and even though some of Fludd's detractors at court maintained that the brothers must have helped him write his volumes, he swears that he has never knowingly seen or met any of them, and that he had finished the treatises of the *Macrocosm* history four or five years before he heard of the Fraternity. As for his detractors, they were made to admit that they had not even read his books, and those who had didn't understand them!

The heart of 'A Philosophical Key' is the tract 'Of the Excellency of Wheat'. This alchemical experiment Fludd said was the basis of his entire philosophy, and is probably the *Tractatus de Tritico* mentioned by Michael Maier and the '*Lib. de Tritico*' listed in the commonplace book of Joachim Morsius, who made a trip to England from October 1619 to February 1620. It was not published until 1623 in the *Anatomiae Amphitheatrum*, and bore the title *De exacta alimenti panis seu tritici anatomia*. It was by then stripped of the Preface, in which Fludd provides insights about the Rosicrucians and the controversy at the court of James I, and includes the same letters from the Continent as in the 'Declaratio'; the 'Calumniator's Vision,' an allegorical and mythological account of the origin of the world; and the Epilogue. Without question, the discovery of 'A Philosophical Key' adds crucial information about Fludd. [26]

Also in 1619, Johannes Kepler published his *Harmonices mundi* (Harmonics of the World), which spelled out his ideas of the harmonics of the universe (see Chapter 6). Kepler was an ardent follower of Plato, Pythagoras and the Neoplatonist Proclus, and believed, in concert with them, that creation had proceeded according to geometric principles. In an intuitive flash in 1595, the 24-year-old mathematics teacher thought he saw how it all fitted together: the sun was in the centre of the universe, as Copernicus theorized, and the orbits of each of the planets fitted exactly into the five symmetrical geometric solids (which is all there are), which was why there were only six planets. His quest, for the rest of his life, was to prove it mathematically, and along the way he made some monumental discoveries, most notably the three laws of planetary motion. While residing in Linz as the provincial mathematician in service to Emperor Rudolf II, he acquired Tycho Brahe's astronomical observations, and in 1618 he finished his *Harmonices mundi*. Before it was printed, however, he saw a copy of Fludd's *Macrocosm*, and added an Appendix taking issue with Fludd's mystical approach as opposed to his mathematical one (see Chapter 5).

Fludd replied to this in 1621, but first, in 1620, he published the second section of the first tract of the *Microcosm* history, and turned his attention to the steel-making issue. He petitioned James I for a patent, which was referred to the Privy Council in May:

> Their Lordships did this day, accordinge to his Majestys gracious pleasure signifyed unto them, take consideracion of a peticion presented to his Majestie by Robert Fludd, doctor of phisick, wherein hee complayneth that havinge at his great expense and charge drawn over hither from forraine parts certaine persons expert and skillfull in makinge of steele, with purpose to imploy and sett them on worke here in that mystery, for the benefitt and behoofe of the publicke, hee is in some hazard and dainger to bee hindered in that his good intent by a new patent nowe sought for by some such as have alreadie discovered their insufficiencie and abuse in that kinde.
>
> Forasmuch as it is conceaved that the makeinge of good and serviceable steele within the kingdome is a matter of good consequence, and whorthie of all due encouragement; their Lordships did intreate the Earl of Arundell, the Lord Digbie and Mr. Chancellor of the Exchequer, or anie two or them, to take notice of the foresaid peticion, and the reasons thereunto annexed, and upon consideration thereof to report their opinion to the Board that such further order may be taken therein as the cause shall require. [27]

The report being favourable, the Council approved Fludd's petition for a patent. It is of sufficient importance to quote the record in full:

At Hampton Court, the 27 of September, 1620.

Present: The Kings most excellent Majestie, Lord Archbishop of Canterburie, Lord Chancellor [Francis Bacon], Lord Stewarde, Lord Admirall, Lord Marquis Hamilton, Lord Chamberlein, Earl of Arundell, Earl of Southampton, Earl of Kellie, Lord Viscount Doncaster, Lord Digbie, Mr. Treasurer, Mr. Comptroller, Mr. Secretarie Naunton, Mr. Secretarie Calvert, Mr. Chancellor of the Exchequer, Master of the Rolls, Master of the Wardes.

Whereas the business concerninge the makeinge of steele within the kingdom hath been of late often debated at the Board, and the patent formerly graunted to certaine persons for the sole makeinge of steele and prohibitinge the importacion of anie from forraine partes recalled and taken in, aswell in respect of the insufficiencie and badness of steele made by those patentees as for other reasons of state expressed in former orders of the Board;

and forasmuch as Robert Fludd, doctor of phisick, hath, at his great charge, drawen over hither from forraine parts certaine persons, and amongst others one John Rochier, a Frenchman, skillfull and expert in makeinge of steele, and, being an humble suitour for a patent to sett them on work in that mystery for the good of the publicke, did this day offerr to the Board a certificate, under the haunds of manie cutlers, blacksmyths, locksmyths and other artificers workinge in steele, that the steele made by the said Rochier is very serviceable, good and sufficient, and Doctor Fludd further undertakeinge to make as good steele as anie is made in forraine parts, and to vent the same at easier and cheaper rates than the outlandish steele;

that they will waste noe wood but only make it of pitt coale; that they desire no barr of importacion more than what the goodnes and cheapnes of their stuffe shall occasion, and thereby that his Majestie shall have a third part of the profitt arryseinge thereby; his Majestie and their Lordships, upon consideracion thereof, findinge it very requisite to have the makeing of good and serviceable steele settled within the kingdome, did well approve of the offers made by Doctor Fludd herein, and thinke fitt that letters patents bee graunted unto him and the said John Rochier upon securitie to be first given aswell for the makeinge

of good and serviceable steele as for aunsweareinge the third part
of the profitt to his Majestie.[28]

The Council issued a warrant to draw up Fludd's patent on
17 November 1620.[29] Exactly what he did with it is not
presently known. But Fludd's position at this point is clear:
he enjoyed the favour of the King subsequent to the discussion
of his philosophy with him; he was an established member of
the medical community; he was welcomed in the scholarly
circles around Sir Robert Bruce Cotton and Bishop
Thornborough; his alchemical laboratory graduated to the
point where he employed technicians to make better steel than
had been previously known in England; and the Privy Council
agreed that his steel was to be preferred above that made by
the prior patentees.

The Rosicrucian business, the slanders at court and the
steel-making patent behind him, in 1621 Fludd returned to
writing and published another part of the unfinished *Microcosm*
history, which issued from de Bry in Frankfurt. De Bry also
published his reply to Kepler in the *Veritatis Proscenium*.
The following year, Kepler continued the debate with his
Pro suo opere harmonices mundi apologia (see Chapter 5 and
Bibliography).

Continuing the grand *Microcosm* scheme, Fludd published
his *Anatomiae Amphitheatrum* (Mystical Anatomy), in 1623,
which was dedicated to John Thornborough, Bishop of
Worcester. Included at the end was his final revision of his
harmonic scheme of the universe and last reply to Kepler, the
Monochordum Mundi, dated 1621.

No sooner had he settled this debate to his satisfaction,
than two new ones appeared on the scene, one major and one
minor. In 1623 Fr Marin Mersenne (1588-1648), the French
mechanist, viciously attacked Fludd's philosophy in his
Quaestiones celeberrimae in Genesim. Mersenne is best known for
the wide correspondence he carried on with many leading
scientific figures of the time and for his unceasing promotion
of the mechanistic world-view. Among other things, he called
Fludd an 'evil magician, a doctor and propagator of foul and

horrendous magic, a heretical magician'. Fludd was appalled by the intemperance of the attack, and thought Mersenne to be mad. He did not reply, however, until 1629, in the *Sophie cum moria certamen*. He also wrote some revealing statements about the affair in *Dr Fludds Answer unto M. Foster* of 1631 (see Chapter 7).

Also in 1623, a small work was published in London under the title *Tillage of Light* by Patrick Scot, Esq, a little-known figure. Scot maintained that the Philosopher's Stone or elixir was allegorical, not material, and therefore useless and foolish to pursue. Fludd was moved to take issue with this position, and wrote an unpublished manuscript entitled 'Truth's Golden Harrow', now in the Bodleian Library, Oxford. Even though Fludd disdained the pursuit of the worldly elixir by calling it *chymia vulgaris*, he nonetheless defended its actual existence (see Chapter 6).

Still working on the uncompleted *Microcosm* history, in 1626 Fludd published his *Philosophia sacra et vere Christiana seu Meteorologia Cosmica* in Frankfurt. It was dedicated to John Williams, Bishop of Lincoln, and was the last part of the *Microcosm* to appear. Whether he wrote additional parts is not known, but eventually Fludd turned his attention from the *Microcosm* to working on his reply to Mersenne and his multi-volume *Medicina Catholica*. It was not until three years later, however, in 1629, that he published his response to Mersenne in the *Sophie cum moria certamen*. Bound with it is a short work by a Joachim Frizius, *Summum Bonum*, which also refutes Mersenne and defends the Rosicrucians. Since the latter is rather Fluddean in tone and deals with the same issues, many, both at the time and later, assumed it to be Fludd's, but he clearly denied being the author. There is evidence that it was he who sent it to the printer (who did not bind it separately, as Fludd requested) after editing, however, and he did say that Frizius' position represented his own.[30]

The first part of the *Medicina Catholica* also came out in 1629. It dealt with the causes of disease, the winds and airborne spirits, and was dedicated to Sir William Paddy, a medical colleague of Fludd's in the College of Physicians.

In the same year another interesting event took place in Fludd's life: he and his heirs were given a grant by Charles I for some service performed for the Crown:

> June 8 Westminister: Grant to Robert Fludd, Doctor of Physic, of a messuage [dwelling-house with associated outbuildings and property] and land in Kirton, Co. Suffolk, come to the Crown by being devised by Richard Smart to Anne Deletto, an alien, to the use of Rosamund Hewett to hold the same to such uses as by the will of Smart are limited. Signified to bee your Ma:ts pleasure by Sir Sidney Mountagu.
>
>   R. Heath. [31]

Exactly what this service was is presently unknown. Perhaps Fludd used some of his continental contacts to gather or pass information, or there may have been some sort of medical treatment involved.

The second part of Fludd's *Medicina Catholica*, which bore the title *Pulsus*, was published about 1630. It dealt with the pulse, and was the first printed work to agree with William Harvey's theory of the circulation of the blood. The same year, Pierre Gassendi, whose help Mersenne had enlisted in 1628 in the polemic against Fludd, published his *Epistolica exercitatio in qua principia philosophiae Roberti Fluddi, medici, releguntur, et ad recentes illius libros adversus R. P. F. Marinum Mersennum . . . respondetur.* Although Gassendi did not agree with Fludd's basic views, he was critical of Mersenne's reckless accusations and found Fludd to be a learned and Christian gentleman (see Chapter 7).

In 1631, the third and fourth parts of the *Medicina Catholica* were published: the *Integrum Morborum Mysterium*, which deals with contraction and rarefaction and is dedicated to George Abbot, Archbishop of Canterbury; and the *KATHOLIKON MEDICORUM KATOPTRON*, which includes diagnosis by astrology and examination of the urine, and is dedicated to Sir Robert Bruce Cotton.

As if to ensure that the Mersenne affair and Fludd's controversies would continue, in 1631 a new combatant entered the lists: a little-known country cleric named William Foster wrote a small tract called *Hoplocrisma Spongus, or a Sponge to wipe away the weapon-salve*, and even went so far as to tack up a notice of the book on Fludd's door. Foster denounced Fludd's advocacy of the weapon-salve (which worked from a distance by a preparation applied to the weapon causing the wound) as using diabolical forces to effect the cure, and backed his argument by referring to Mersenne's attack on Fludd as an evil magician. The incensed physician dictated a quick reply in English, *Dr Fludds Answer unto M. Foster, or the squeesing of Parson Fosters Sponge, ordained by him for the wiping away of the weapon-salve*, which provides some interesting information about Fludd, his publisher, his relationship with James I, and the Mersenne-Gassendi affair, and also dispatches Foster's contentions to the extent that he was never heard from again (see Chapter 7).

Then Fludd prepared a final reply to Mersenne and Gassendi that was published in 1633, the *Clavis Philosophiae et Alchymiae Fluddannae*, which once again defended his philosophy in characteristic terms. That same year he was once again elected a Censor of the College of Physicians. The next year, he was sworn in as a brother in the Barber-Surgeons' Company, presumably to do surgery (which physicians had done by barber-surgeons) without criticism.

On 8 September 1637, Fludd died in his house in London. According to his wishes, specified in his will, his undisturbed body was wrapped in fresh linen and taken to the inn at Bearsted in Kent to wait for nightfall. Then a torchlight procession proceeded to the burial site at the Bearsted parish church. Following interment, all those at the funeral (many of whom Fludd had himself provided with their funereal dress) returned to the inn and were entertained at his expense.[32]

In the floor of the church, a brass plate over his grave reads:

In Iesu qui mihi omnia in vita morte resurgam
Under this stone resteth the body of Robert Fludd Doctor of Phisicke who changed this transitory life for an imortall

the VIII day of September A.D. MDCXXXVII
being LXIII years of age, whose monument
is erected in this chancell according
to the forme by him prescribed

As stated on the brass plate above, Fludd's nephew did have
a monument made for him as specified in the will. It has Fludd
in front of an open book that rests on top of a tall pedestal,
which gives the impression of a lectern. Slightly up and behind
his hatless head is his coat of arms, which are centred under
an arched alcove. On the face of the pedestal is the inscription:

Sacred to the Memory

of the Illustrious Physician and Man Robert Fludd, alias *De
Fluctibus*, Doctor of both Faculties, who after some years of
travelling beyond seas, undertaken successfully for the
improvement of his mind, was at length restored to his Fatherland
and was not undeservedly received into the Society of the London
College of Physicians. He exchanged life for death peacefully on
the 8th day of the month of September A° D^mi MDCXXXVII,
in the 63rd year of his age.[33]

No costly perfumes from this urn ascend;
In gorgeous tomb thine ashes do not lie;
The mortal part alone to earth we give;
The records of thy mind can never die;
For he who writes like thee — though dead —
Erects a tomb that last for aye.[34]

Thomas Fludd of Gore Court, Otham, in Kent, Esquire, erected
this Monument to the happy Memory of his most dear Uncle on
the 10th day of the month of August, MDCXXXVIII.[35]

On either side of the inscription are arranged small marble
blocks with the names of seven of Fludd's books, as he
prescribed them: *Phisica et Technica Macrocosmi Historia*; *Phisica
et Technica Microcosmi Historia*; *Misterium Cabalisticum*;
Amphitheatrum Anatomicum; *Philosophia Sacra*; *Misterium*

Sanitatis and *Misterium Morborum.*[36]

At the time of his death, Fludd had been at work on another long volume explaining his philosophy, which was published posthumously in Latin (Gouda, 1638) and English (London, 1659). Bearing the title *The Mosaicall Philosophy: Grounded upon the Essential Truth or Eternal Sapience*, his last work runs (in some 300 folio pages) through the basic tenants of his philosophy, namely, that it is grounded in the true wisdom of God, as expressed esoterically by Moses in his books and elsewhere in the Bible; that alchemical and weather-glass experiments can prove his theories about the order and nature of the universe; that the universe was created in numerical, geometric, musically harmonic and a cabalistically consistent way; that of the ancients, only Plato and Hermes can claim the title divine, because they alone understood the Mosaic principles; that the philosophy of Aristotle is animal, terrene, and diabolical; and that experiments with the loadstone by Gilbert and Ridley prove, contrary to Aristotle, that there is action at a distance in nature (see Chapter 8).

In this book, written at the close of his active and productive life, Fludd wanted to make one main point that went to the heart of his philosophy, no matter how complex in detail, and that was also the basis for his answer to his critics. The work clearly shows him to be a Renaissance Christian Neoplatonist, and the greatest summarizer of that tradition of his age:

> . . . it is most apparent that there is no art or science, whether it be abstruse and mystical, or manifestly known, be it speculative or practical, but had his root and beginning from this true wisdom, without that act and virtue whereof, no true and essential learning and knowledge can be gotten in this world, but all will prove bastardly or spuriously begotten, having their foundation not upon Christ, the true ground, firm rock, and stable cornerstone, on which all verity is erected, for as much as only in Him is the plenitude of divinity; but placing the basis or foundation of their knowledge upon the prestigious sands of

imagination; namely, after the inventions or traditions of men, and according unto the elements of this world, from whence they gather the fruits of their worldly or human wisdom, that is quite opposite in effect unto the true wisdom; namely, the eternal one, which hath his root and original from God, and not from man.[37]

Notes

General note: Spelling and punctuation in Chapters 4, 6, 7 and 8 have been modernized where possible to improve readability.

1. See William H. Huffman, *Robert Fludd and the End of the Renaissance* (London and New York, Routledge, 1988), pp.4ff.
2. Ibid.
3. Ibid.
4. Ibid.
5. Ibid.
6. Ibid.
7. Ibid., pp.14ff.
8. Ibid.
9. Ibid.
10. Historical Manuscripts Commission, Great Britain, Report 9, Salisbury MSS XViii, pp.241-2.
11. Fludd says in 'A Philosophical Key' (see Chapter 4) that he finished his *Macrocosm* history four or five years before he heard of the Brothers of the Rosy Cross, therefore about 1610 or 1611.
12. Walter Pagel, *William Harvey's Biological Ideas* (Basel and New York: S. Karger, 1975).
13. Allen Debus, *The English Paracelsians* (New York: Franklin Watts, 1965), pp.104-5.
14. For more detail about the circles, see Huffman, pp.25-35.
15. Huffman, pp.51-3, 135ff.
16. F. N. Pryce, 'Introduction' *The Fame and Confession of the Fraternity R.C.:* by Eugenius Philathes [Thomas Vaughan]. (Privately Printed, 1923; reprint, 1988), p.12.
17. Same edition as above, p.32 of Vaughan's edition of 1652. See Huffman, pp.137ff.
18. Huffman, pp.144ff.
19. Ibid., pp.145-6.

20. Ibid., pp.146–8.
21. Ibid.
22. For a discussion of the Rosicrucians and Fludd, see Huffman, 'The Rosicrucian Connection', pp.135–66.
23. *Acts of the Privy Council of England, January 1681-June 1619*, London: HMSO, 1929, p.135; Huffman, pp.23–4.
24. Joachim Morsius, who made a trip to England from October 1619 to February 1620, listed among Robert Fludd's works a *Lib. de Tritico*. See Huffman, p.154.
25. See Huffman, 'Royal Patronage', pp.36–49.
26. Ibid., pp.43–5; Allen Debus, *Robert Fludd and His Philosophical Key* (New York: Science History Publications, 1979).
27. *Acts of the Privy Council of England, July 1619-June 1621* (London: HMSO, 1930, p.212).
28. Ibid., pp.284–5.
29. Ibid., p.319.
30. Huffman, pp.157–8.
31. State Papers 39/27 29.
32. Public Record Office, Probate II/175; the entire of the test is found in Huffman, pp.222–9.
33. Translation by A. E. Waite, *The Brotherhood of the Rosy Cross* (1924; reprint, New Hyde Park, N.Y.: University Books, 1961), p.279.
34. Ibid.
35. Ibid.
36. The names are as spelled out in his will: Public Record Office, Probate II/175; see Huffman, p.222.
37. Fludd, *Mosaicall Philosophy*, p.25.

Chronology

1574	*January.* Born at Milgate House, Bearsted, Kent. Baptized in Bearsted Parish Church, 17 January.
1592	*25 January.* Mother, Elizabeth Andrews Fludd, dies at Milgate House.
	10 November. Enters St John's College, Oxford.
1596	*3 February.* Receives BA from St John's.
1596–8	Studies for MA at St John's. Writes treatise on music.
1598	*8 July.* Receives MA from St John's, says he is going overseas.
1598–1604	Travels in France, Spain, Italy and Germany. Tutors Duke of Guise and his brother. Writes treatises on arithmetic, geometry, perspective, military arts, the art of memory, geomancy, motion, astrology.
1604 or 1605	Enters Christ Church, Oxford.
1605	*16 May.* Receives MB and MD, licenced to practise medicine.
	8 November. First examination by College of Physicians in London to practise medicine there.
1606	*7 February.* Examined by College, given permission to practise.
	2 May. Questioned by College about allegations of arrogance concerning supremacy of chemical

medicines over Galenical.

July. Travels to France to confer with colleagues from Italy and France.

1607   *30 May.* Father, Sir Thomas Fludd, dies at Milgate House.

1 August, 9 October, 22 December. Further examined by the College.

1608   *21 March.* Offends Censors of College of Physicians by examination replies. Candidacy for Fellowship in College revoked.

25 June. Readmitted as candidate for Fellowship in College of Physicians.

1609   *20 September.* Admitted as a Fellow of College of Physicians of London.

c.1610   Completes manuscript of *History of the Macrocosm.* Read by John Selden, medical colleague Dr Richard Andrewes and others.

1614   John Selden praises Fludd's medical skill in his *Titles of Honour.*

Fama Fraternitatis of the Order of the Rosy Cross published in Germany.

1615   *Confessio Fraternitatis R.C.* published in Germany.

Andreas Libavius attacks the Fraternity in *Analysis Confessionis Fraternitatis de Rosia Cruce.*

1616   Fludd replies to Libavius with *Apologia Compendiaria.* This brief Apology includes an outline for a longer work and a letter to the Fraternity.

1617   The longer defence, *Tractatus apologeticus integritatem Societatis de Rosea Cruce defendens*, published in Leiden by Basson.

The *Tractatus Theologo-Philosophicus . . . de Vita, Morte et Resurrectione* published in Oppenheim by de Bry. Also from the de Bry press appeared the first part of Fludd's *magnum opus: Utriusque cosmi majoris scilicet et minoris metaphysica, physica, atque technica historia* (Technical, Physical and Metaphysical History of the Macrocosm and Microcosm). Contains Volume I,

History of the Macrocosm, Tractate I. Dedicated to God and James I.

1618   *De Naturae Simia* (The Ape of Nature) printed by de Bry in Oppenheim. Tractate II of the *History of the Macrocosm*.

12 May. Two holders of the monopoly patent to make steel in England complain to the Privy Council that Fludd is making steel illegally in his chemical laboratory.

Fludd elected a Censor of the College of Physicians. Called before James I to defend his Apology and Macrocosm History.

Interview gains Fludd the favour and patronage of the King.

Writes 'Declaratio Brevis' at the suggestion of James.

1619   Volume II, *History of the Microcosm*, Tractate I, published by de Bry in Oppenheim: *Tomus Secundus, de supernaturali, naturali, praeternaturali et contranaturali microcosmi historia*.

Fludd writes 'A Philosophical Key' as sequel to the 'Declaratio Brevis'.

Johannes Kepler publishes his *Harmonices mundi*. Includes a lengthy appendix attacking Fludd's version of the Neoplatonic harmonies of the universe.

1620   *30 May*. James I charges Privy Council to consider Fludd's petition to be granted a patent to make steel.

27 June. Fludd gives public anatomy lecture at College of Physicians.

27 September. The Privy Council grants Fludd a patent to make steel after considering testimony of its superiority.

1621   Another part of the unfinished Microcosm History published by de Bry in Frankfurt: *Tomi secundi tractatus secundus, de praeternaturali utriusque mundi historia*.

The *Veritatis Proscenium* also published in Frankfurt by de Bry.

Replies to Kepler's attack.

1622   Kepler replies to Fludd in his *Pro suo opere harmonices mundi apologia*.

1623   *Anatomiae Amphitheatrum*, Fludd's mystical anatomy, published in Frankfurt by de Bry. Dedicated to John Thornborough, Bishop of Worcester. At the end is the *Monochordum Mundi*, dated 1621, Fludd's final revised universal harmonic scheme and last reply to Kepler. Marin Mersenne, the French mechanist, attacks Fludd's mystical philosophy and science in his *Quaestioned celeberrimae in Genesim*.

1625   Death of James I, accession of Charles I.

1626   *Philosophia sacra et vere Christiana seu Meteorologia Cosmica* published in Frankfurt by Officina Bryana. Dedicated to John Williams, Bishop of Lincoln.

1627   Elected a Censor of the College of Physicians.

20 July. Inspects alum works with William Harvey and six others from the College at the order of the Privy Council.

1629   *8 June*. Grant by Charles I to Fludd and his heirs of a 'messuage and lands' in Suffolk.

Replies to attacks by Mersenne in *Sophie cum moria certamen*.

Bound with it is the *Summum Bonum*, dealing with the Rosicrucians, by Joachim Frizius.

First part of the *Medicina Catholica* published by Fitzer in Frankfurt.

Dedicated to Sir William Paddy.

1630(?)   *Pulsus*, the second part of the *Medicina Catholica*, published in Frankfurt, the first printed work to agree with William Harvey's circulation of the blood theory.

Pierre Gassendi publishes his examination of Fludd's works, done at Mersenne's request, in his *Epistolica*.

Donates twenty-four books to Jesus College, Oxford.

1631   The third part of the Catholic Medicine, *Integrum*

Morborum Mysterium, dedicated to George Abbot, Archbishop of Canterbury, and the fourth part, *KATHOLIKON MEDICORUM KATOPTRON*, dedicated to Sir Robert Bruce Cotton, published in Frankfurt by William Fitzer.

William Foster, an English cleric, attacks Fludd's views on the weapon-salve as diabolical. Fludd replies with *Doctor Fludds Answer unto M. Foster.*

1633   Elected a Censor of the College of Physicians.

The *Clavis Philosophiae Et Alchymiae Fluddanae*, the final refutation of Mersenne and Gassendi, published in Frankfurt by Fitzer.

1634   *12 June.* Sworn in as a brother in the Barber-Surgeon's Company.

Elected a Censor of the College of Physicians.

1637   *8 September.* Dies in his house in London.

Buried in Bearsted Parish Church.

Leaves Manuscript published as *Philosophia Moysaica* in Gouda, 1638, and *Mosaicall Philosophy*, London, 1659.

Rosicrucian Defence:
Apologia Compendiaria[1]

A Brief Apology,
washing away and cleansing the stain
of suspicion and infamy applied to
the Fraternity of the Rosy Cross
with, as it were, a Fludd of truth.

*An abbreviated apology, cleansing the stigma attached to
the life and dignity of the Fraternity of the Rosy Cross.*

When I had carefully considered that the misery of this life
would only be ended by blissful death, I observed that human
wandering on earth should be compared . . . to the restless sea,
which, when churned up with agitated waves, causes incessant
danger to the corporeal ship propelled here and there on an
uncertain course, and causes the sailor . . . to be utterly
ignorant of how he might by chance bring his ship to an
unknown port of good fortune. Thus it happens that hardly
one out of a thousand directs his life to that longed-for goal
of happiness.

Yet the wisdom of the ancients bears witness to the fact that there is a certain sure and undoubted seat of human happiness in this world which some auspiciously investigate with long wandering and due inquiry; and by this wisdom we are taught that Moses, the master of divine philosophy, who certainly conversed with God and obtained the key to both types of understanding (supernatural and natural) by divine assistance and illumination of the most Holy Spirit, reached the shores of happiness. Moses' virtue was attained by Bezaleel, Joshua, David, Solomon and all the Prophets, whose wisdom, in fact, some of the ancient philosophers imitated. Among these, Mercurius Trismegistus especially assumed and laid claim to a place for himself — his admirable understanding of things above and below has been depicted vividly for us by his Sacred Sermons and his knowledge in the Smaragdine Table.

Certainly the praises of Apollonius the Pythagorean have not been buried or overwhelmed either by the forgetfulness or silence of men (as happens so frequently) because of Philostratus, by whose authority it was published that Apollonius understood all arcane things, called back to life the dead, cured the sick, was pre-eminent in his own sanctity, and abounded in every happiness of this life. From Philostratus we have also learned that a fraternity of Wise Men flourished at that time in India; and he attributed them with the ability to predict the future events of men, to read and foresee all their conditions, parents, fortunes and names, as if these things were engraved on their brow in the arcane characters of Nature. The wisdom of both the Egyptians and Ethiopians must be entrusted to the eternal monuments of memory. From the founts and springs of their wisdom, Plato is said to have drawn his divine understanding of the Idea. Beyond doubt all these (if they were immune to Diabolical Superstitions) reached the summit and regal peak of mundane happiness and wisdom. [2]

Therefore, as a result of both frequent pondering and concern about these things, I perceived that men must seek nothing more eagerly than happiness; and that they must contemplate nothing more seriously than true wisdom, inasmuch as this has revealed to men a life that is long and free from all poverty. They must also have a contempt for the world and seek the divine contemplation of natural mysteries as well as an understanding of supernatural mysteries; and seek the pure revelation and admiration of the divine majesty.

For indeed, as a result of my pondering I am wracked with great sadness because the Wise Men of old so sparingly and covetously sowed the tree and fruit of their understanding that they are no use at all to our generation . . . For under the cover of words divine secrets lurk and are hidden. But since the Lord God is kind and merciful, He has promised that He will lay open the doors of His understanding and revelation to those of us seeking entrance. Therefore let us strive not to quench the spirit, not to reject prophecies, but test everything; and uphold what is good (I Thes. 5).

The *Fama* of the Society of the Rosy Cross has traversed almost all the provinces of Europe and at last has reached our ears; it demonstrates an admirable knowledge in both divine and natural secrets. Therefore why should this order not be diligently investigated and, once discovered, approached, since it has been disclosed in writing and spoken word that [the members] will offer themselves to us willingly, of their own accord, without compulsion, and without compensation. And why should these gifts of wisdom offered to us be rejected so lightly when (as Hippocrates attests) life is short, the art of living long, and opportunity fleeting?

To which in a dubious manner D. Libavius is evidently responding in his *Analysis of the Confession of the Fraternity of the Rosy Cross.*[3] As a result of a careful examination of his writings, we detect more resentment and maliciousness than inquiry into the root of the matter. For in one place, he has brought the brothers into suspicion of sedition. Elsewhere he contends that their promises cannot be accomplished or achieved without abhorrent Magic or Diabolical illusion. Moreover, he is eager to defile them with heresy, very likely by contaminating the truth with falsehood, since, as the utmost injustice, it is customarily held that he accuses someone of something and charges another with vice and evil, as it were, by the effects, before the cause is accurately explored and adequately examined.

It is because of this, therefore, that I penetrate into the purpose of the fraternity, and have proposed to defend it rationally, both with weapons peculiar to itself, and with those drawn from the quiver of others against the accusation[s] of D. Libavius and aspersions of others. Let us listen (I implore) to the brothers speaking on their own behalf, in their confession, in which, as I believe, they have, in the eyes of the world, absolved themselves of those very things of which D. Libavius accused them.

They have quite certainly satisfied the wise, for they acknowledge that it is Gods plan to extoll the humble and restrain the prideful with contempt. Wherefore all suspicion of sedition must be removed from them (if they are speaking openly and willingly), since the mother of sedition is ambition, which pride is wont to raise and nourish. They also acknowledge that they embrace the divine life and delight in the contemplation of God and all of Nature in its nakedness; and they hold the world in contempt, since this is a valley of miseries, in which there is no daily fortune and no bliss before the end. Indeed, by this confession (unless I am deceived), they have cleared themselves of this charge, and have removed from themselves all fancy of sedition.

Elsewhere in the confession, we find that they purely and sincerely embrace Christ and live the Christian life: wherefore, all suspicion of Diabolical art must be far removed from them and banished to the farthest shores of oblivion. For those who truly and sincerely revere Christ are able to complete and happily bring to success greater things by His Sacred virtue rather than by the deceitful and empty illusions of the Devil, of which we conclude the brothers to be utterly free.

In one place of their confession they relate that they have been devoted to the true philosophy and given to the wisdom of the Gymnosophists; and in another place, that this notion of theirs has not proceeded from the impulse of their own will, but from the counsel of the Holy Spirit, whose Nature it is to teach not with deceptions, but with the deepest truth. From this we perceive that this society not only observes the pure and exposed laws of Nature in the arcane matters set forth; but also . . . they acknowledge the help, not of the malignant spirit, but the divine and holy one, without whose divine power the true key of understanding cannot be attained.

For why (I ask) would the brothers, in exalted language and with the blare (as it were) of a trumpet, divulge to the world their admirable beneficence and stroke the ears of men with empty rumours if they should be unable to demonstrate the credibility of their promises without the wicked cunning of Magic, since in every Christian realm it has been decreed and ordained that Necromancers, sorcerers and conjurers should be punished by the death penalty, pillory or fire.

But since D. Libavius is evidently fixed in this belief and opinion that astounding things of this kind are not possible without Diabolical assistance, we have resolved to publish a more extensive Apology by the next market day [i.e. the Frankfurt Book Fair], in which we will declare that it is likely that the published purpose of the brothers shines out to the world and is brought into view in part by divine gift and in part by the arcane knowledge and light that has been hidden and secretly preserved in the viscera of Nature.

After he had explored all the hiding places of suspicion, D. Libavius noticed that in the end the brothers had fallen into heresy inasmuch as they promised to reform the failures of the arts and to remedy and correct their weaknesses. And yet we see in the most preeminent Theological scholarship that the context of the Holy Bible is a constant matter of controversy, since one school of theologians interprets something in this way, whereas another interprets it that way, and all are accustomed to twist and explain in their own way any of the more difficult places in the Scriptures, and this occurs without any suspicion of heresy, although it is the custom and habit of Romans and Papists to reproach violently those who do not adhere firmly to their religion. Thus Lutherans, Calvinists and others of this kind, as well as the brothers [of the Rosy Cross], must be regarded as heretics if any credence can be given to the empty and threatening assertions of the Papists.

Therefore, after I had carefully and rationally thought these things over, I perceived that carelessness in man must never be so ignorantly punished with the lash so as to cause one to spurn and reject lightly such gifts of wisdom offered to us from, perhaps, the divine will, since it is the Nature of a wise man to accept with a grateful mind those things which are good and to reject wisely the bad as being unworthy. For prophesying must not be rejected, but all must be tried and what things are good must be upheld (I Thes. 5). We are also exhorted to pursue love and to grasp not spiritual gifts so that we may prophesy (I Corinth. 14). The propositions of the brothers, (I say) must not be rejected, since it is possible that a true enlightening of the Holy Spirit and a full abundance of understanding in fact is appropriately distributed unto individual men of this age by God, as accordingly were the Prophets (Joel 2). God promised, through the mouth of his Prophets, that he wished his own spirit to pour forth over all flesh; that the sons and daughters of men will prophesy; and that the young men will have visions and the old men will have dreams. Similarly, He promised to all who would ask in His name He would give the Holy Spirit, which would teach them the whole truth (Luke 12; John 14, 15). And elsewhere all will be taught by God (Isai. 54; Jerm. 31; John 6), from whom alone, and not from anyone else, must the true key of wisdom and understanding be obtained. It is not [true] that even the Pharisees, although unworthy, obtained this key of understanding (Luke 11).

Yet he [Libavius] forbids these things to men of this age, and if any enjoy them by the auspicious will [of God], he immediately accuses them of Necromancy and superstitious Magic. Although it is wicked and vile to attack someone without due inquiry into the case and when suspicion is not strongly evident, he is quick to condemn.

Therefore let us inquire into and weigh carefully the actions of the brothers, and let us consider with a keen and attentive mind their character and circumstances. And after all things have been carefully thought out, let us grasp with a just mind, and let us picture to ourselves, whether these brothers are from God or the Devil. We find that the Holy Spirit is known from its gifts. Therefore let us consider whether or not these gifts are found in the brotherhood. One reads that in each individual the Holy Spirit shines forth for the advantage of all (I Corint. 12). Thus Moses, Aron and the other Patriarchs and Prophets received the Holy Spirit, not for their own sake, but in order that they might teach, exhort, admonish, rebuke, console and direct the people into the fear of God. For this reason they publicly disclosed their spiritual and secular gifts. Therefore the evidence for the enjoyment of the Holy Spirit in the brothers is not small, since they have published (in fact) their prophesies and knowledge for the benefit of all. Yet it must be investigated further and examined more deeply whether their beneficences and deeds coincide with the other gifts of the spirit.

First of all, let us consider with an open mind what the signs and evidence are from which the presence of the Holy Spirit is made known. Then let us pursue and follow their tracks in the confession, and look deep inside to see how many traces of the aforesaid signs may be detected. Those who speak the truth, prophesy, speak in strange tongues, interpret Scripture, drive out demons, fully care for and heal the sick, observe the divine precepts, do not resist the word of God, are occupied with the fruits of the spirit (i.e. Love, Joy, Peace, Charity, Generosity, Humanity, Goodness, Gentleness, Temperance and Chastity); and those who do not follow the wants of the flesh (examples of which are Fornication, Impurity, Impudence, Idolatry, Sorcery, Hostility, Homicide, Gluttony, Anger, Discord and other things of this kind); these people are without doubt from God and full of His Holy Spirit, since these are the infallible signs by which the sacred Scriptures teach us to recognize men enlightened by the Holy Spirit and distinguish them from others.

As a result of careful examination of the confession of the brothers, carried out step by step, we conclude that they prophesy concerning the future, the duration and the renovation of the world. They openly declare the Nature of all things by means of a strange tongue and arcane writings;

they speak the truth inasmuch as, in our generation, they say, all falsehood ends; they do not resist the word of God, since they live uprightly; they observe the divine precepts, since they write that they do this not from an impulse of their own will, but as a result of the Holy Spirit and the counsels of God; they are generous, for they, on their own accord, give generously to us from their own treasures; they are endowed with humanity, since they do not wish to be received by us, but invite us to their palaces; they are free from idolatry, since they wholly and sincerely embrace Christ; they cure the sick, for they suggest ways the afflicted can be healed; they abound in charity, since they offer to the world all their secrets and treasures, and administer their cures without compensation.

After these things have been carefully considered, we find that much evidence of the Holy Spirit shines forth in the brothers of this community (if their deeds agree in the same tone with their words). Therefore this is the principal reason why I have plunged myself into their cause, whereby we may remove the poorly founded suspicions of D. Libavius and others. Those who ... believe from extremely perverse opinions that these things are deemed either Diabolical or deceptive (which seem novel to hear or strange and unusual to see, or seem arduous and certainly beyond the grasp of the imagination): if they will explore them a little more accurately, they will feel not only that they are open to detection, but also (in fact) easy to do without any demonic art. [This is] just as we will clearly declare in our larger Apology, in which we have proposed the most true and certain ways by which the understanding of spiritual and worldly secrets, and the accomplishment of things commonly thought to be wondrous, may be accomplished: namely, by the outpouring of the Holy Spirit in men, by which means we read that the Apostles performed their miracles; or by the revelation of the mysteries concealed in Nature, which also requires the assistance of the Spirit. It was by this means that Solomon was versed in every kind of knowledge.

It also happens sometimes that the Devil, through his impious Prophets, that is to say, Necromancers, tells the truth, and issues divinations about the future; yet he has not been able to do this without the divine will, without which the formers prophesies are only implanted with the lies of conjecture. Thus the spirit of the Devil poured forth into the world by its own will is either deceptive, false or enigmatic. Thus was that diabolical apparition deceptive that appeared to Brutus, urging him to battle with false persuasions, promising victory, and finally causing his death on the battlefront when he obeyed its counsels. Likewise this unreliable and enigmatic spirit directed that pagan oracle of Apollo and governed the tongue of the raving Pythia. Sometimes this malignant spirit will mix truth with falsehood, since by occasionally announcing the truth, it more easily, at a later date, deceives and leads to inevitable ruin those who have put faith in it by means of its absurd and outrageous lies. He also very often contrives sudden destruction among men, which (in fact) he produces by his own person: thunder, lightning, earthquake, turbulent water, destruction of buildings, disturbance of violent winds and other things of similar effect; or he succeeds through his ministers, i.e. Sorcerers, Cacomagicians, Necromancers, etc.

Therefore, after all these things have been sensibly and rightly rejected, we will find that all the works of the Devil are wicked and pernicious to such an extent that they have produced no true understanding or benefit, but rather (on the contrary) have produced for the world a multitude of gross misfortune, terror, constant danger, and wretched confusion; the cause of which is evident, since he is the avenger of the living God and destroyer of things forsaken by God.

And thus we conclude this short Apology, standing (as it were) as a harbinger and precursor at this spring Book Fair and in lieu of a prelude and proemium will announce the coming of a fuller Apology (which is to appear at the next Book Fair), the title and subject of which you may grasp from the brief appendix.

Apologia

An apology defending the integrity of the Society of the Rosy Cross

In which it is proved, contrary to the accusations of D. Libavius and others of the same kind, that the wondrous things offered to us by the Fraternity of the Rosy Cross are possible to achieve without the wicked deceit of Magic or the deceptions and illusions of the Devil.

Abridgement or short summary of the Apology to be published

We have divided the Apology into three parts:

In the FIRST Part,
> We have expressed the scope of our design. It is discussed whether Magic, Cabala and Astrology are arts or superstition.
> The books of God, both the visible and invisible, and especially that great book of Nature and living things, are treated.
> The kinds of divine Scripture are clearly explained.
> The characters and letters of the same Scripture are examined, and are observed to be formed in a two-fold manner, that is to say, either by the word FIAT in the Creation, or by the Sacred Finger of God after the Creation; and it was in the latter way the Book of Revelation of the divine Majesty was written, as were the laws of Moses on the stone tablets.
> Finally, we have declared that the will of God and His laws can be discovered in the characters of the great book of Nature and in the signs of the stars without any help from the Devil.

In the SECOND part, we have demonstrated:
> That there is a very great defect in almost all the arts which have until now flourished in the schools;
> The causes and origins of the defects of those arts;
> The defects of the Physical arts, i.e. Natural Philosophy,

Medicine, Alchemy, Mathematics, namely Arithmetic, Music, Geometry, Optics and Astrology; of the Moral arts and the arts pertaining to the instruction of men, i.e. Ethics, Economics, Politics, Jurisprudence; and finally the impediments of the Theologians. The remedy and correction and reformation of these failing and declining arts which are not accomplished without the instructive leadership of the uncreated Holy Spirit and his ministers, namely, the good angel, wondrous light, or the wise man or the truth-speaking prophet saying Urim and Thummim, Epos.

In the LAST part, also declared are:

The origin of the universal[ly] created light and mundane spirit, wondrous effects, occult properties, arcane mysteries.

That it is possible that someone may advise and reveal his plans to another who is hundreds of miles away by means of both the living voice and appearance, and letters (and this is accomplished without the operations of the Devil): and this is accomplished both by natural and super-celestial assistance.

That it is possible that someone who writes only a single book may know all the other things which have been described in all books; and this is accomplished by: the admirable virtue of the arcane light, which is the active medium; and the universal spirit, which is the passive medium.

The admirable power of the true and arcane Music in all creation, both animate and inanimate; that the secrets are infinite and wondrous that lurk under the surface and visible form and exterior [of] even the smallest things.

The Author's Epilogue to the Brothers of the Rosy Cross

O Brothers most dear, who promise to renew, with a reciprocal effort, the spiritual forms of men and improve their fortunes, which have been transformed into more miserable images by the fall and misfortunes of Adam, I bid you greeting in the name of Him who you worship wholly and uprightly, Jesus Christ, Saviour of the world.

Behold, I ask for your indulgence if I, an unskilled philosopher and unworthy spokesman of your praise, have stumbled into some evident error and defective knowledge in my little work. Listen briefly to who I am, who indeed would be most willing to do the utmost for the companions of your order, whereby I might satisfy the acquisitive ears of men with a whispering more reliable and worthy of your praise. I am by name as written above, or, if you wish, Fludd, noble enough by birth but of the lowest rank of Medical Doctor at London. My wife is the desire for wisdom, my children the rewards arising therefrom. The body is a prison, and the desires of the world are empty and destructive of the mind. I long to be a mirror of myself whereby I may contemplate what I am. I have traversed and surveyed with my eyes and mind almost all the provinces of Europe: the surging deep seas, the high mountains and slippery valleys, the crudities of villages, the rudeness of towns and the arrogance of cities; the ambition, avarice, infidelity, ignorance, idleness, deception and almost all the miserable aspects of men; and I have found no one correctly knowing himself according to God (John 1). Life was light and this light was in the shadows and the shadows seized it not. Therefore I have detected nothing except vanity triumphant everywhere throughout the world. The vanity of vanities (I say) has been discovered everywhere and all things are misery and vanity. And thus now I close. Farewell and be mindful of me.

FINIS

Notes
1. Translated by Robert A. Seelinger, Jr, and William H. Huffman.
2. Fludd shows here he is an intellectual descendant of Marsilio Ficino and the Florentine Academy in the belief in a pure divine wisdom that was passed down from Moses to selected illuminated initiates over the centuries. See William H.

Huffman, *Robert Fludd and the End of the Renaissance* (Routledge, 1988), pp.87-99; Frances Yates, *Giordano Bruno and the Hermetic Tradition* (Routledge & Kegan Paul, 1964), pp.1-19; D. P. Walker, *The Ancient Theology* (Cornell University Press, 1972), pp.1-21.
3. See Introduction, p.23.

The Technical, Physical and Metaphysical History of the Macrocosm and Microcosm: *Utriusque cosmi . . . Historia*

BOOK ONE

The History of the Macrocosm

CHAPTER ONE

On infinite nature and the maker of all things

Infinite nature, which is boundless Spirit, unutterable, not intelligible, outside of all imagination, beyond all essence, unnameable, known only to the heart, most wise, most merciful, FATHER, WORD, HOLY SPIRIT, the highest and only good, incomprehensible in height, the unity of all creatures, which is stronger than all power, greater than all distinction, more worthy than all praise, indivisible TRINITY, most splendid and indescribable light, in short, the divine mind, free and separate from mortal matter, glory of all, necessity, extremity and renewal: Here, I say, GOD, the highest and greatest of all, whose name was made blessed in eternity, skilfully formed the admirable machine of the entire Macrocosm, and beautifully adorned his structure. Even all the Philosophers, both ancient and more recent, except the

Peripatetics who maintain that the world exists from eternity, had generally preserved, with unanimous consent, this most divine founder of the Universe with many appellations, giving fitting surnames to any quality of his action. Among these Hermes calls him eternal; Thales — the most ancient; Plato — the innate source of the universe; others — the infinite cause, both outside all things and yet in all things and everywhere; others — the Being of Beings, the Prime Cause insofar as other causes are derived from it, the maker and originator of all, and according to Plato, the repository of understanding. For this reason, some others have called him the Prince of worldly things, the ruler, the arranger, the first mover and the first moving, the immovable beginning: because all things are moved by him, though he remains immovable, and not subject to suffering, since he is always stable, fixed and permanent; Trithemius calls him the source of all worlds, by that he seems to understand the sole maker of the creation. Next, Plato and Mercurius Trismegistus call him father, in so far as he is the originator of all fecundity and the begetter of all things; finally, the greater chorus of the Philosophers (among whom are named Democritus and Orpheus) concludes that GOD contains every name, since all is in him and he himself is in everything, not unlike the manner in which all straight lines drawn from the centre to the circumference are said to be in the centre: or just as Number is said to exist in unity, which is the common measure, source and origin of all numbers, and contains every number, joined to itself in a unique way. Wherefore, the Pythagoreans, and the most learned in science of numbers, likened the Monad or Unity to God the Maker, because he was, alone and by himself, before all, and also because he was the first mover and only activator that existed to complete the huge structure of the creation; moreover, they attributed the Dyad or Duality to his matter, or subject that he worked on, because it was the second constituent upon which the Maker worked to complete the world; and finally they assigned the Triad to that spiritual power, or fiery, shining essence of his, by which the said matter or substrate of the world was brought from the potential state to the actual, and all parts and regions of the fabric of creation and the creature therein clearly distinguished and divided, because it was this very essence that took on the role of the third constituent entity in the make-up of the whole creation.

It is clear from the opinions of these pagan Philosophers that even they did not stray far from a true knowledge of the divine mind: among whom I can never sufficiently praise nor admire that sacred revelation of Mercurius Trismegistus, and his profundity concerning the divine mysteries, since his *Pymander*, of divine and superhuman derivation, bares the hidden secrets of God and his whole creation to us. We reckon that Plato was the next man who knew most about the divine, because, contrary to the assertions of the Peripatetics, he constantly maintained that the world and all that is therein, whether visible or invisible, is made from the innate and essential first principle or material. However impossible it may be correctly to describe the true essence of this triple Creator, nevertheless, following certain learned mens ideas, he is depicted thus, namely, in human form.

The reasons why some have depicted God in human form are because

Man, made in the image of God is his representative, for he contains Mind, Word and Spirit (Trismegistus in *Pymander*).

The Father of all, existing as Mind, Life, and Light, created man like himself, just as he made his Son, with whom he was well pleased because he was beautiful, in the image of the Father (ibid).

Man is made of Life and Light; God is made of Life and Light (ibid).

God made man in his own image (Genesis I).

Or like the Sun; the Pagans and the Orientals in particular, worship God in this form:

Wise men have correctly compared the Deity to the Sun, because

All spiritual and invisible brightness comes from God and all things are inwardly illuminated by the same, just as all visible, external light and brightness come from the Sun; for as the Sun is Emperor of material things, God, holding a fiery sceptre, is Emperor of spiritual things.

According to the wise Sidrac, three things are observed in the Sun's nature, namely the Sun itself, its shining light, and its life-giving heat; they liken the Sun to God the Father himself, its brightness to the Son, and its warmth to the Holy Ghost. Brightness is born and derived directly from the body of the Sun, the effect of heat and its kindly power come from the other.

Or like a triangle, because three separate angles are found in a single thing, just as three persons exist in the unity of God, and are in no way separate from that unity. And it seems to us that the incomprehensibility and infinite extent of the divine Triangle may quite well be brought within the range of human powers of understanding by [a] diagram [see Figures 1 and 2, p.96 and p.97].

God can be compared to this incomprehensible Triangle, for it contains everything in itself, extends everywhere with its power, drives everything, and there is nothing outside it; furthermore, the corner made by any angle corresponds to one third of the world, nevertheless, all are joined in one unique object, and constitute one, single, regular, equilateral and perfect triangle; and at the same time it is clear from this diagram that, just as all the corners of the above Triangle sticking out of the spherical surface of the world are found to be similar and equal in form, dimension and size, so also the three persons in the oneness of the divine essence are equal to each other in their degree of perfection ...

CHAPTER FOUR

A description of the Primal Matter, or raw material of the Maker of all

The matter of subject, in the middle of which the great creator of the Macrocosm laid out his construction, is the Philosophical Hyle, that Physicists have called the absolutely Primal Matter; though there is nothing commoner than it, there is nothing that is less understood. Two of the ancient philosophers in particular have zealously studied and investigated it; of whom the first, namely Thales (with whom Heraclitus and Hesiod agree), after long study and much burning of the midnight oil, firmly stated that *Water was the primal matter of Nature*, and gave a good reason for his opinion. For, said he, water is the seed and food of all natural things. Both Pagans and the writers of Holy Scripture seem to approve of this notion ...

Aristotle himself says that the difficulty of understanding this matter is that it cannot be understood in isolation, nor described by itself alone, but only by analogy, that is by examples drawn from other things ... The godly Moses, therefore, compared it now to the earth, without form and void, now to the waters and the abyss; Mercurius Trismegistus, in his *Pymander*, to a dreadful shadow, turning into a watery substance; Plato to the mother, nurse, and house of things that are born, because it contains and nourishes everything; Augustine ... to darkness and silence; Pythagoras to a duality; some to a mirror in which the soul of the world is seen ... And the reason why Plato called it infinite is because it is inclined to take endless forms ...

We therefore conclude from all the writings of the Philosophers, ancient and modern, that this primal material is a primordial, infinite, shapeless Existence, as suitable for something as for nothing; having no size or dimension, for it cannot be said to be either large or small; having no qualities, for it is neither thin, nor thick, nor perceptible; having no properties nor tendencies, neither moving nor still, without colour, or any elementary property. However, it is the original passive ground of action, containing the world; therefore it is called the mother of the world, whose bosom embraces, like a mothers womb, the heavenly spheres, adorned with golden fires, and the four elements, hanging lower near the centre . . .

CHAPTER SIX

The Universal Essence, with which the Creator of the entire creation shaped Matter

The most wise Creator of the World, who said, 'I am the light of the world, the true light, and the father of lights', has decreed that this shapeless matter, this foundation of nature, or underlying ground of the fabric of the universe, should be the dwelling-house of Forms, and (here I agree with Plato) their nurse, so that the mass buried in darkness throughout the abyss can be made visible and perceptible by their presence, stirred into action and smoothly finished; and He has most generously united the brilliance of His light, first to the Empyrean sky [heaven] (for *the breath of His mouth*, which Mercurius Trismegistus calls the light of God, and the divine power of His spirit, was *carried over the waters*) then, secondarily, to the Ethereal sky [heaven], and the Sun, and the rest of His creatures in the heavenly spheres, so that they may adorn the Ethereal sky by means of their power, and imbue lower creatures with shape and life.

Moreover, this fiery creation, made on the first day in the divine image, this first, best gift given by God to perfect the remainder of his construction, Moses called Light . . . Plato the Idea, Aristotle, the First Principle . . . Mercurius Trismegistus in his divine discourse has called this same light the blessed brilliance that blossomed forth in the beginning, and asserts that the elements were made out of wetness and dryness; Marsilius Ficino, in his *Commentary* upon that discourse, said that the formless folds of matter were illuminated with shape by that blessed brilliance. In short, all of them with almost complete unanimity, usually called it the first Action, Pattern, Species and Essence. This light, therefore, is a unique material, that has substance, that exists unmixed, and is the purest, worthiest, and noblest of all; so great is its nobility indeed (see Augustine on Genesis), that the more matter partakes of the nature of light, the more perfect and noble it is therefore considered to be . . .

. . . doubtless it is from this supercelestial spring of light that the invisible fire, which Zoraster and Heraclitus tell us is the mother of all, is derived: and its power, though invisible, seems to be felt and acknowledged by every living being and the rest of sublunary creation, down to the sea itself and its daily and nightly ebb and flow. Furthermore, the Philosophers call this fire sheer Action, because of its vigour; nothing is more rarefied, nor more penetrating because of its fineness, nothing is more useful, nor filled with more power, nothing is more excellent nor more uniform in nature, there is no (greater or lesser) improving and finishing agent for succeeding Existences, nor does anything wreak things up faster, nor more easily loosen their connecting parts . . .

We therefore conclude that light is *either* uncreated, that is, it is God the Creator of all (for the true light is in God the Father himself, his brilliance shines and overflows in his Son, and his brightness burns in the Holy Spirit, passing all understanding), or it is created from that Uncreated which is the purest soul, the true natural pattern and the clearest spirit of the three heavens, as well as being their boundary and their vehicle for giving them shape. Therefore, Iamblicus thought that because the Ether was so bright it was nothing but light itself, of which all creatures are composed; for indeed angels are infused with a certain shining intelligence, which is poured into all reasoning objects, but in differing amounts according to the ability of their natures to accept it. Next it descends to the celestials, where it produces a life-giving power in them, and by means of this life-giving brilliance the lower creatures are clothed with life, and the ability to breed.

In men it is the process of clear reasoning, in the other animals it is the hidden fire that evidently governs the actions of their life and senses; in vegetables it is a sort of bright soul, that hides round their central parts and causes them to grow and multiply endlessly; in minerals it is the spark of brilliance impelling them to their goal of perfection. The macrocosms of the skies [heavens], and also the differences between the bodies that exist in them, arise from that supernatural light created on the first day; for one finds, indeed, that the highest heaven differs from the lowest, and the lowest from the middle according to the excess or deficiency of this essence in them, and that any element of each sky [heaven] is pushed or raised up by its absence or presence, for the further Matter is from nobleness of Natural Pattern, the grosser, more impure, darker and less worthy it is. Hence, indeed, arises the diversity of created objects and substances, hence too their perfection and imperfection, their rawness and ripeness, volatility and fixity, thickness and rarefaction, darkness and brilliance, weight and lightness, and the proportions that distinguish one thing from another . . .

CHAPTER SEVEN

The two unmixed primary elements derived from the first two substances; also the two elements arising from these unmixed ones

From each of the two aforesaid parts of the composite there arises a certain pure, unmixed element that displays the nature of its origin. For when the fiery spirit was borne upon the waters . . . it produced a certain inherent character as an accompaniment to its actions. The Philosophers have therefore said that this driving force is as the effect of pattern, or enduring Light, and have also called it Heat; by its nature it penetrates density to the centre, opens up dense and solid things, by penetrating, breaking them down scatters and separates their thickness from their thinness, cleanses them by rarefying and thinning them, and makes them volatile and light, so that they are naturally raised and carried up on high. After this, it collects homogeneous things together, digests, matures, fixes and perfects them; while it cuts heterogeneous things open, disperses, pounds and hammers them. The reason it perfects everything is because it is directly derived from the worthiest and most rarefied body of all, and closely accompanies its most fine and noble substance.

Having described the origin of this active element, it remains to reveal the properties or elementary nature of the other constituent part — this is passive, since action can do nothing without passivity at its disposal. Therefore, such an arrangement must be derived from the Hyle or Primal Matter of the Macrocosm by a disturbance in the darkness, before the beginnings of Heat. For, with the first appearance of Light, the shadows began to move, and as Heat was produced by the motion of Light, so another element, tolerant of the activity of Heat, and therefore called passive, because it suffers, rather than acts, was produced by the motion of the substance of the abyss; it accompanies the dense substance of matter, and directs it or lets it relax, just as Heat does the substance of Light, which is seen to be struggling, or relaxed, according to whether there is much or little of it. Once Heat arose, the element of Cold already present became totally opposed and inimical to hotness, and withstood the inroads of its subtle penetration, and by resisting, and at the same time collecting the impure, dark, dense part of the abyss' substance, constricted, condensed and contracted it towards the centre of the cosmos, which is the place to which Cold naturally escapes, because it is furthest from the outside rim of the world [universe], which is where the great pure essence of the Spirit of Light existed and had its place from the beginning.

Hence comes the solidity of the elements of the lower heaven, the density, thickness, and darkness of the earth. This is also the reason that no weight can pass beyond its central point even if a hole were bored straight though the whole mass of the earth, because the central point of the earth is the part furthest from the circumference of the universe, and therefore the spot where Cold naturally desires to come to rest, where it is free and immune from the enmity of Heat and its hostile action. We therefore conclude that Heat likes to embrace formative ability, confers on everything the functions of life, and is its souls companion: whereas Cold resides in material bodies, cherishes them, and is naturally inherent in them. As a result of the mutual activity and passivity of these two, certain parts of the Macrocosm are more or less rarefied, due to the recession of Cold and the advance of Heat — these are the upper parts — and certain become dense, due to the absence of Heat and the presence of Cold — these are the lower ones. So the world [universe] is made and its regions marked out . . .

Now two other systems arise from these pure and simple elements or qualities, which Artefius called Elements of the Elements, and they are Wetness and Dryness: the first is made by mixing and tempering the two simple elements. For if Heat is joined to an equal amount of Cold, the substance Wetness is produced: the quality connected with Heat or Cold in the complete absence of moisture we call Dryness. It is certain, therefore, that Dryness in nothing but the absence of Wetness, and therefore these two qualities are said to be contrary to each other; still, they are passive with respect to the primary elements . . .

CHAPTER NINE

The universal substance of the [heavens] . . .

After completion of the divine creation, therefore, a portion of the substance of each and every region was discovered to be either an element, or made of elements; the part called 'element' is the original, smallest and purest part, and the part of each sky [heaven] made of elements is formed directly from it. And since the substance of each and every heaven is denser, dirtier and less perfect, according to whether more or less light is present in it, its elemental simplicity may be thought of as absolute [in the case of the spiritual or Empyrean Heaven] . . .

or relative. For if the elements [of the highest heaven] are compared to those of the lower . . . they are called pure and unalloyed, but with respect to [the lower in comparison with] the upper . . . [heaven] they are judged thick, composed of matter, and therefore less pure.

For the sacred scribes admit that the elements of the spiritual sky [heaven], because of their mysterious simpleness, spiritual purity, and indescribable powers, do not exist only in the station frequented by blessed Intelligences and angels, but are also contained in GOD himself, the arranger and supporter of the entire structure. For there is in the angels a stability of nature, and an earthy strength, which are the sturdy seats of GOD; in the angels there is also mercy and piety, which are a watery and cleansing power; in them also there is a tenuous, airy Spirit, which is why they are called the wings or feathers of the wind in Holy Scripture; and finally, there is in them love, which is a most splendid fire. Therefore the royal Psalmist says, 'Thou hast made thine angels spirits, and thy ministers a flame of fire.' Seraphim, Virtues and Powers are, moreover, fiery; Cherubim, terrestrial; Thrones and Archangels, aquatic; Dominions and Principalities, aerial. Finally, it is said of the Archetype himself: 'Let the earth be opened and bring forth a Saviour.' . . . Likewise, the sphere of the Trinity is doubtless the brighter and more immaterial part of the spiritual heaven for it is to the realm of the Trinity that the Seraphim are nearest. Its royal throne or base is the firmament or crystal sphere, which St John called the glassy sea, in which he noted something earthy, something rarefied, and something watery; when the extremes have been carefully considered, the intermediates are easily understood [see Figure 7, p.173].

Elements are also found throughout the ethereal sky [heaven], of which the greater part is called Quintessence, because it is raised above the four elements of the lower . . . [heaven], in worth, in immateriality and in position. For in this region there is the solidity of earth without the viscosity of water; and the liveliness of air without its flowing fire and a non-combustible, incorruptible, preservative, bright heat, putting life into the bodies in the lower . . . [heaven] with its warmth, because it is directly derived from the spiritual . . . [heaven]. The things created in this region . . . also share in its elements, because both the fixed and wandering stars claim their high rank from having these elements. Thus, the stars that are called fiery, because they are a union of fires heat and its dryness, are Mars and the Sun among the planets, and Aries and Leo among the fixed constellations; those that are called airy, because they have warm, wet natures, are Jupiter and Venus, and Gemini and Libra; Saturn and Capricorn are called earthy because of their melancholy nature; next, Luna is credited with a watery nature, because she is named the Empress of Moisture by the Philosophers, and the same applies to the signs of Cancer and Pisces in the Zodiac. This is why astrologers first observed the threefold nature of this heaven in the beginning, and without this observation, they could not have perfected their art. It is also via the properties of these elements that any planet produces its varying effects here below . . .

Finally, no one, not even an ignoramus, can be unaware that the lowest space in the universe is occupied by elements and things created from elements; but here they are more impure, grosser, and more subject to our senses, and more liable to decay, and they are far less noble and less pure and simple than in the other higher regions. And for the same reason it is recognized that created objects composed of them take their characteristics from a proper mixture of them. So the cold, wet, hot or dry characters of vegetables and minerals are described and distinguished according to which elements are predominant in them. Elements are therefore found everywhere, and in all regions of the universe, but no one can be thoroughly aware of the various regions for the scarcity or abundance of universal pattern in them, and why elements are cloudy and thick in the lower region, in the middle one purer and clearer, and at the top, or in the supercelestial region, highly spiritual and blessed in every way . . .

CHAPTER TEN

Chaos, and the origins of the creatures of the lowest heaven

. . . the Primal Matter of the universe is multiple: that which we call *Hyle* is furthest from the compositeness of the world; *second* is *the material of the elements*, which may be distinguished because it is ruled by either Heat or Cold . . . *Third* is *Chaos*, the confused matter and undigested mass, into which you may think all the four elements being shuffled and jumbled; for earth and water, heavier than the rest, rose to the orbit of the moon, while fire and air, though lighter, descended to the centre of the earth . . . And finally this (Chaos) is the material that ancient Philosophers, modern Chemists and Poets have sung and written of in their books: they maintain that it produces the diverse essences necessary at one time for the completion of the Macrocosm, and in another, for the completion of the Philosophic Work. The *fourth* and final Primal Matter to which the Philosophers have given a name is that closest to solid bodies, which directly generates the compound creatures of the lowest world, and causes them to multiply: we recognize sperm in animals, seed in plants, and *argent vive* in minerals as being of this kind . . . And, following and idea common to many Philosophers, we also describe Chaos as the place where Cold and Wet fight with Heat and Dryness . . .

BOOK TWO

The Structure of the Cosmos

Contents of this Second Book

THE HIGHEST
in which two
things are to
be noted, its

Three Parts,
or heavens.

Of the Trinity,
boundless (Ch. 2, 3).
The Empyrean,
(Ch. 3).
The Crystalline,
(Ch. 4).

Its constitution,
made up of

The purest
formative light
(Ch. 5).
The purest Spirit,
ever new and
incomprehensible.

THE MIDDLE,
Called Ether,
Which has
(Ch. 6, 7, 8)
THE WORLD
has three
regions
namely,

Eight parts, the
houses of the
stars.

Fixed
Wandering

Its constitution,
made up of,

Spirit, neither too
gross, nor too fine
Light that is half-
way to matter

Three parts

THE EXTREMES:
The upper, the
tabernacle of Fire
(Ch. 11).
The lower, the
residence of Earth
(Ch. 12).
The middle, called
the sphere of
Moisture, divided
into the regions

THE LOWEST,
in which two

are to be
obseı ved
(Ch. 10)

of
AIR (Ch. 13)
WATER

It constitution,
whose two parts
are

TERTIARY Light,
the grossest of
all (Ch. 14 & 15)
SPIRIT, the
thickest and
cloudiest of all.

CHAPTER ONE

How the universe is divided

... we are about to deal with the double world — the Macrocosm and the Microcosm — distinguishing the first from Man, or the Microcosm, and considering the entire volume of the Primal Matter as the World, Cosmos, or Macrocosm, which the spiritual light, or spirit, of the Lord, surrounding the waters, enfolds in its circular embrace. A certain round portion of the abyss' material is divided into three different parts by Formative Nature, by arranging light and darkness in varying blends, i.e. distinguishing between the blends according to their purity and impurity. The highest of these parts is that expanse of the universe where the fiery spirit was, and it is contained in the primal light-stuff, extending from the concave limit of the Sphere of the Trinity to the convex surface of the Starry Sky. And because it contains an abundance of pattern, the matter of this part of the universe is so fine and pure that it is totally imperceptible, and cannot be seen by us; so the Philosophers called it intellectual, and spiritual in the highest degree. Now the middle part is adorned with the stars, both fixed and wandering, for it occupies the entire cavity of the Universe held between the convex outline of the Moons orbit, and the concave outline of the *Primum Mobile*; and the matter in this region is solid compared to the higher one. Finally the lowest part of the universe is the entire expanse embraced by the concave outline of the Moons orbit. It therefore follows that the mass of the universe is divided into two main parts, of which one is bodiless, spiritual, and very pure and fine, that is the upper one, whilst the other one is material. And this material part is subdivided again into two parts: one is rare, fine and imperishable — that is the middle part — the other, gross, impure, and subject to corruption — this is the bottom part of the universe, which we shall call sublunary and elemental.

These, then are those regions of the Macrocosm that both ancient Philosophers and the authors of Holy Writ called the Heavens. Now they called the highest, Empyrean, fiery Heaven, because it is filled throughout with spiritual fire, or the material of light. And this is the Third Heaven into which St Paul the Divine said he was snatched by the Spirit, and he called it Paradise; nevertheless, all our writers customarily called the middle region the Ethereal Heaven, and Heaven itself. Finally, Matthew the Apostle seems to call the whole airy mass between Ether and Earth, 'Heaven'. . . . And since rarefying earth makes water, rarefying water makes air, and rarefying that makes fire (for one element rotates into another), one must say that the whole of this region may be called Heaven; for St Peter call it the material of firmamental Heaven and Earth. Indeed, Ezekiel thus described the nature and situation of these three Heavens in his vision, Chapter 1 . . . It is therefore clear that all the matter in the world is nothing but these three Heavens, differentiated according to thickness and thinness, or purity and impurity.

CHAPTER THREE

The first three days of creation

. . . it is probable that before there was any creation the divine spirit was sent forth and was carried over the waters, and went round about them, making three round trips, and that each of these round trips produced a spiritual and supernatural day. So it constituted the first day by its first circuit, in which is finished the Empyrean Heaven, and pushed the darkness down one step. In its second revolution it brought forth the second day, during which darkness was pushed down another step by the doubled amount of divine power, and so light and divine power shone out in the middle, ethereal region of the universe. Finally, during its third and last revolution, darkness was thrust down a third step to its final distance from the circumference by virtue of the third application of the divine. Thus the tabernacle of the one world was completed, and all its regions, its powers and its properties produced by the power of the Holy Spirit. Here these three days, measured only by the action of the Holy Spirit, come to an end, recalling the number of the persons in the Trinity, all of which were active in the creation of the world . . .

CHAPTER FOUR

The creation, nature and operation of light, and the constitution of the Empyrean Heaven . . .

The Lord said, 'Let there be light', and the highest heaven, overflowing with light, was created on the first day; for the contents of this [heaven] are nothing but the most essential and spiritual light-matter, fabricated from a large amount of Soul and a little highly rarefied Spirit; therefore it is called the fiery spirit, and the Empyrean [heaven], fiery and burning: this, as Isadore says, is not because it burns like the element fire, but because it is steeped throughout in the splendour and brightness of light . . .

So the mist and darkness of this hitherto shapeless and obscured region were driven down from above in a spiral turn by the presence of this fiery force of creation . . . Therefore, this heaven, or highest supernatural region of the world, is the abode of the fountain of light, a sort of fatherland, or native land, in which primordial light was first created and located. Therefore, it is called the foundation of the world, and because its presence produces form, it satisfies the desires of all the matter in this [heaven] fully: for it is certain that the matter which is in this heaven is very fine and thin, since it is wholly and essentially concerned with producing shapes; for according to Aristotle (*Metaphysics* 4) wherever there is more Pattern, there is less Matter, consequently solidity of Matter means scarcity of Pattern; for many a shining pattern diligently rarefies by its activities the purest part of the Abyss' matter, retained in this [heaven] by the Divine Will, and causes its substance to expand, as is seen in the art of Chemistry, where, if some gross matter, shut up in a fairly large, hermetically sealed flask, is made fine by a strong heat, its rarefied portions naturally want a bigger space, so the glass breaks, so that the rarefied vapours may find more room.

In the same way, the transparent matter of this heaven, rarefied by its friendly warmth, spreads itself out widely. This is why (according to Basilius, *Hexameron*, Book 2) this particular heaven expands itself among all the skies.

We therefore conclude that this heaven is the noblest of all in its form, quite unimaginable, spiritual and pure, greater in quantity than all the others because its matter has been thinned out, in quality brightest and fullest of fiery force, in action most active, its light-stuff moves the fastest, the heat derived from its motion is the gentlest and most moderate, friendly, being neither burning nor destructive, but connected to all lower things by properly proportioned bonds, in a peaceful pact, and these it conserves within its own spirit, which can hardly be imagined on account of its thinness; it lives and moves of itself, without attracting any foreign bodies, unimpeded by the cold which naturally co-exists in the same spot, and it has a spherical shape which it acquired both from the motion of the Spirit borne over the water, and from the power and brilliance proceeding from that Spirit by the Divine Will. It is called by name Empyrean, or Fiery (see above). It is also called Olympus from its clearness. It is called supernatural too, for it surpasses natural laws in being spiritual rather than substantial, and intellectual, because we cannot see it, for it is intelligible and perceptible only to argument *à posteriori*, or else because it is filled up with angels and Intelligences.

CHAPTER SIX

The middle heaven, its various names, and the reasons for them

The middle region of the universe, created on the second day, has various names because of the action of the light-stuff as it extended downwards; for, taken by itself, with regard to its own particular material, it is called the Middle Spirit, after the dispersal of darkness: compared to the upper sky, that is, to light-stuff, or mixture of light-stuff and spirit, it is called Ether: However, in comparison with the four elements that appeared and were separated on the third day, it is called Quintessence: by itself, in its own substance as revealed on the second day, it displays the forms both of highly rarefied water, and of waters mist vapours, never still. For this reason the Platonists called it the Middle Spirit of the world; in Chapter 5, Book 1, above, we called it the container, binder and vehicle of the Soul of the World, where we demonstrated the reason for this name. Because this spirit is the band around the light descending from the fountain of spirit, by whose action a natural and vivifying heat is stirred up, it is called Ether, that is Shining, for according to Isadore, it is perpetually bright and clear; or it is called fiery, according to Anaxagoras, who said that all that burns should be called Ether, for the Ether is nothing but the air on fire, which, in Greek grammar, is derived from *aitho*, which means 'I burn'; the spiritual atmosphere is, as it were, 'ath-ear', that is, burning spirit . . .

Finally, it is a single element, just as Aristotle says in his book *Meteoron*, where he asserts that the Ether is nothing but a single element. So this single element must be additional in number to the four lower elements soon to be produced in the lowest region; hence the Philosophers call it Quintessence . . .

CHAPTER SEVEN

The composition of the Middle Sky

As the Empyrean was created on the first day out of a single thing, namely light-stuff, it was made up of three-quarters of the thinnest fire, and one-quarter of the most highly rarified spirit; therefore, it is completely formative, compared to the lower regions, and because fire predominates in it, it is called Fiery. So the matter in this second sky, since it is more solid than that of the upper, primal one, is matched to its shining pattern in the proper proportion, so that its matter and its form are of equal weight in the balance; hence it is composed of one part of tenuous and fiery light, descending out of the fountain of the Empyrean Sky through the membrane of the Crystalline Sphere, and an equal quantity of the still unformed, powerful, Primal Matter, called Spirit. We also see Pythagoras assert this when he writes that the Moon, the Sun, and the rest of the stars are composed of fire and air. This is also why Artefius, in agreement with the ancient Philosophers, called this region the Sky of Equality, in whose centre, learned men located the region of the soul. From this it seems likely that this sky is made up of two quarters of the Middle Spirit and the same quantity of that very beautiful matter, whose shining soul is prepared to satisfy the desire of its spiritual material, and preserve it from change and decay . . .

The final reason is a musical one. One finds that Matter and Light in this middle region are in the proportions of one-and-a-half to one, so that when Matter ascends upward, and Pattern descends downward, in this Middle Sky, or Ether, they produce the harmonious sound of a diapente (the interval of a fifth) . . . [See Figures 6, 7 and 8, pp.172, 173, 174-5.]

CHAPTER EIGHT

The recoil of the double darkness of the Middle Sky into the lowest region of the universe . . .

The power and brightness of the supernatural sky, descending at the Creators command through the openings, or rather teats of the watery sieve, and pouring down as if from the perennial spring scattered with its presences the dark fogs of this portion of matter, and drove them against the lowest region, hitherto concealed and invisible, and thus made its matter visible, formed, pure, fine and rare, and much better in perfection and beauty than the substance of the lower sky [heaven]. It is, indeed, that bright essence gushing in endless numbers of small streams from the original supercelestial spring into every part of that celestial sea, as the ancients called the Ether . . . And this light, although secondary, is so full of active and vivifying strength, that it frees the whole of this sky from its original uncleanness, and drives its darkness away, making it glide in a spiral down to the lowest part of the universe, like a dreadful gloom . . . In a word, it is the secondary essence that moved about for three days after the creation of the world, instead of the Sun, and on the fourth day, as Dionysius says, it became the substance of the Sun; it is indeed action, pure and simple, by means of which this Ethereal Sky is most skilfully adorned with sparkling fires; and for various reasons there arose its various motions, from east to west, and according to latitude, called oscillations; and finally and most necessarily a kindly heat was kindled, friendly to the life of created things . . .

. . . matter in this heaven is denser than that in the Empyrean Sky, therefore light works more strongly in it, and it resists in equal measure, and therefore produces Heat as an effect, to drive away the injurious Cold, which is by nature the invariable companion of gross matter, and formerly concealed, is now revealed by the presence of Heat. As this Heat is more intense than that of the highest heaven, likewise it is much more moderate, fine and simple than that of the lower heaven, for it follows the pattern of its mold, from whose activity it has just resulted.

We therefore conclude that, just as the matter that is in the form of this sky is born of an immortal parent, and is incorruptible, worthy, exalted and ennobled by its origin, breathing life and soul into worldly creatures and cherishing them with its Light, so also the Heat derived from its activity and mobility is vivifying, natural, sweet, gentle, moderate and incorruptible; the innate warmth of animals, vegetables and minerals is drunk from this fountain, and its blessed arrival begets, enlivens, increases and brings to perfection all lower bodies . . . It is especially necessary for the continued existence of living beings that light should somehow shine on them at night. For according to the doctrine of Hippocrates, if the life of the stars did not temper the thickness of the air at night, their bodies and their spirits would suffocate; and yet this influence of the stars by night is not noticed, or rather their glory, scattered everywhere, is felt hardly, or with difficulty, except by the eyes.

CHAPTER TEN

The third heaven and its various names. 'Why is this bottom region called Heaven?'

Before the action of light on the third and lowest region of the world, its contents were a sort of gloom, or excessively thick, dark smoke, not endowed with any visible shape, but marked out to some extent by the density of its material, which is thicker towards the bottom, because of the presence of light and heat. Now after the third day, as the influence of light was inclining downward toward the creation of the earth round a mathematical point, this sky was, I say, completed, and as mentioned below, Philosophers both ancient and modern applied various titles to it; for some wishing to signify its size, called it 'the sublunary region of the universe', meaning the entire space measured out between the concave surface of the middle heaven, or Sphere of the Moon, and the convex surface of the physical centre, or earth. Others, thinking of the circuits of the elements, whose substance fills this heaven to capacity, have called it the *region of the elements*, although we have previously declared that the upper elements also fill the skies in some way; they called it 'subject to decay' because its matter is inclined and disposed to desire new forms and relinquish the old.

However, it appears that it may be heaven for two reasons: for it is a portion of heaven spoken of by the general name of 'heaven' dividing the earth from the sky; and also, in various places in Holy Scripture, birds are called the flying creatures of heaven, where the word 'heaven' is meant in a narrow, and not a general, sense. This second title, therefore, makes it certain that there are four different gradations in this heaven, distinguished from each other by the presence of a greater or lesser degree of third-order formative light, making them thinner, more, active, livelier and lighter, or thicker, slower, stiller and heavier; for the further Matter is from the nobleness of Pattern, the grosser, impurer, less worthy and more obscured it is — witness Aristotle in his book *De Substantiis*, so this is why the four elements, differing in their habitat, nature, worthiness and perfection, arise. This is the true explanation of the Hermetic riddle, stating that nothing is higher unless it is lower, not thick unless it is thin, and the chosen matter of the world is the same unique substance throughout. And doubtless, if the light created on the first day by the Divine Will were put out, a Philosopher would probably observe the material of the world return to the same state, for all substances differ from each other in their form only, and as long as they exist in the one condition, so that no part of the Hyle is grander, thinner, thicker or lighter than another, nor can it be considered out of tune with the rest because of some different property.

CHAPTER FIFTEEN

What the aforesaid elements are made of

. . . lastly, the spirit of the darkness in the third and lowest region of the universe, which is by far the densest and heaviest of all, succeeds the spirit of Ether in the natural order, since it is its close neighbour; and, therefore, it occupies the region of Water, and is called Water, but the fragments of this spirit that contain any portion of the light peculiar to this heaven produce minerals. For this reason the retreating spirits of darkness, disturbed from the Empyrean Sky and the highest region of all, composed the element Fire, and when thrown down out of the middle region, or Ether, they escaped and became the element Air; and finally, when they were expelled from the lowest region of the sky, they acquired the essential nature of Water.

Now Earth is like the midden of them all, the receptacle of their surpluses, and (here I agree with the chemical philosophers), the *caput mortum* or dung of the whole spiritual mass; next to its grossness and filth is Water, which is, as it were, the more servile part of Air, as Air is the more servile part of Fire, and Fire of the Quintessence.

Now the following is the chemical extraction of a substance: the better part is the Quintessence, which is raised to the Ether along with its soul, as we discover when we rectify spirits of wine: the grosser part of the spirit, called burning water, is distilled next, in which there are, as it were, three parts of Quintessence and one of phlegma, likened to the denser part of the spirit of darkness in the middle heaven constituting the element Fire. After this [it] follows that weak liquid commonly called *Aqua Vitae*, in which there are two parts of spirits or wine, and the same number of water, which is compared to the element Air, and finally phlegma is distilled, in which there is almost no spirit, but is all a watery mucus, in which the element Water is collected. Moreover, the dregs or *caput mortum* of all of these, found at the bottom of the flask, may not improperly be likened to the element Earth. The marvellous way in which Nature works, selecting and purifying the material of the elements and putting them in order is thus clearly demonstrated . . .

Robert Fludd's 'Declaratio Brevis'
to James I

A brief Declaration
dedicated to the most serene and powerful Prince
and Lord, Lord JAMES, King of
Great Britain, France and Ireland,
Defender of the Faith

In which the sincere intention of a certain publication
is explained very clearly to Your Royal Majesty
by the present author, Robert Fludd,
Esquire and Doctor of Medicine,
the most loyal subject of
Your Royal Majesty

Behold, Most Serene King, that I have ... brought Your
Majestys auspicious mandate to its utmost conclusion: I have
composed, on your gracious suggestion, A Declaration, not
an Apology ... [and] pray beforehand for forgiveness if I ...
am guilty of any troublesome error or unpolished style. For
who, understanding the depth of Your Majestys judgement,
would not be disinclined to dedicate his books or writings to
him [if they contained errors] or [subject them] to his scrutiny,

by whose eyes ... he examines every corner of a writer or speaker, and immediately discovers those errors in the works of others which are not ... perceived by the authors themselves.

But in fact, Your Majesty orders, and my sense of obedience urges that something be published for the eyes of Your Majesty, whereby the face of truth will be disclosed with the veil of doubt removed, and thereby may obtain the most desired thanks of Your Majesty. But my zeal towards Your Majesty urges me to pulsate soothingly the receptive ears of Your Majesty with the calm whispering of truth.

First of all, I will begin to unfold to Your Majesty the line of reasoning of my published *Tractatus Apologeticus* so that henceforth all fancy of religious innovation or suspicion of heresy may be banished forever, and that the just intention of my mind in the publication of this work may be more easily and clearly known to Your Majesty. Therefore, in the first place, Your Majesty, it does not deal with religious innovation, nor does it share even an iota of any heresy, inasmuch as I, the author of that work, have steadfastly adhered to this reformed religion (which is now the custom among us) from my infancy, and indeed almost from the time I lay at the breast of my nurse in England at the very beginning of my life and right up to this day; and I acknowledge and confess in the presence of God and Your Majesty, from the very bottom of my heart, that I remain a perfectly chaste man.

Next, it is well known that my Apology, defending the Brothers of the Rosy Cross against the attacks of D. Libavius, first of all pertains particularly to the impediments of the Arts, which are in a state of decline, and the method of reviving them; and then afterwards considers the wondrous qualities of Art and Nature with philosophical argument and by frequently using the affirmations of the ancients. Furthermore, this school of Philosophers is acknowledged even by the Germans, Catholics as well as Lutherans (among whom the Brothers are said to live), to embrace firmly the Calvinist Religion, just as it is rightful to infer from these lines extracted from a certain letter sent by a friend of mine from Frankfurt:

The Fraternity of the Rosy Cross has been attacked by so many German writers, and clearly it is rejected by them, while they think the Theosophy of the Brothers is the Theology of the Calvinists, etc.

Your most loyal friend,
Justus Helt[1]

Besides, even the Brothers themselves confirm in their Confession that they profess the reformed religion of Germany. From this, therefore, it is sufficiently evident to Your Majesty that neither the motive of innovating religion, nor the affection for some heresy influenced me to publish the *Tractatus Apologeticus* and to have it distributed publicly. Consequently, concerning the reason that I published this *Tractatus* for the favour of the Fraternity, I will openly set forth to the just eyes of Your Majesty the integrity of my heart and the truth of the matter itself, and with the number of words reduced to a minimum, I will clearly show it here in such a way that Your Majesty will not have any further doubts about my faithfulness towards God and Your Majesty and to the fatherland.

There are two spurs exciting me to an understanding of a familiarity with these celebrated philosophers: the revelation of the true basis of natural philosophy, commonly unknown to this day, and the discovery of the profound secret of medicine, celebrated so much with praises by Theography and Philosophy; this Society of the Rosy Cross professes to possess these two gifts of God and Nature (His servant) in the following propositions:

Proposition I

The true philosophy, commonly thought of as new, which destroys the old, is the head, the sum, the foundation, and the embracer of all Disciplines, Sciences, and Arts. This true philosophy, if we contemplate our world, will contain much of Theology and Medicine, but little of Jurisprudence; it will diligently investigate heaven and earth, and will sufficiently, by its images, explore, examine, and depict Man, who is unique.

Proposition II

We are able to show certain modest truths and things that are useful to our country by which its various illnesses can be cured. These truths are not to be divulged in a common manner which is uncertain and inconstant, but in a new way, unknown to the world, which is most certain and infallible.

And so these companion propositions have been the two lamps for my mind, which is most avid for knowledge, and as it were, have been those two blessed luminaries Castor and Pollux, by whose sparks a great desire has been kindled in me to comprehend these men who are most eminent in character and in their promises. And indeed (in my opinion) these propositions are sufficient for any curious spirit worthy of being inflamed by the fire of understanding, which spirit in times past has experienced the leprous notion of Philosophy and Medicine and their inconstancy.

Moreover, if Your Majesty will deem it worthy to read through the lives of all the most distinguished philosophers and most skilled physicians (namely among the philosophers, Plato, Pythagoras, Thales of Melissus, Aristotle, Anaxagoras, Empedocles, Orpheus, Apollonius of Tyana, Hermes Trismegistus, and many others) you will surely discover that all of them, for the sake of learning and erudition and to become participants in the mysteries of the divine philosophy, made long and laborious journeys through almost all of the learned world, in the manner of pilgrims, for the purpose of visiting the wise men of Ethiopia, examining the mysteries of the Egyptian High Priests, and for pursuing the enigmas of the hieroglyphics as well as the secrets of the inscriptions and carvings on the Pyramids of Memphis. And they even journeyed to oriental regions of the world, there to become thoroughly learned in the doctrine of natural things from the Magi or wise men of Babylonia, Persia and India.

Whereby it has come about that by acquiring the immortal title of divinity from some of these sages, they are called Divine even by Christians, since they acknowledged the Trinity not only of God, but also of the uncreated Persons. In this number, Plato and Hermes are counted in particular. Thus also Your Majesty will discover that among the physicians, Apollo, Aesculapius, Chiron the Centaur, Hippocrates of Cos, Chrysippus of Sicily, Aristratus of Macedonia, Euperices of Trinacria, Herosilus of Rhodes, Galen and many other men who were most learned and excellent in medicine also travelled through many nations of the world for the sake of acquiring experience in philosophy and medicine. They traversed all corners of the known world to discover the hidden mysteries of philosophy and medicine by searching and by wearisome exertion; and above all, they visited the celebrated temple of Diana of Ephesus so that they might behold the medical records preserved therein.

Whence at last they had become provided with knowledge, enriched and ennobled from their experience, they returned to their homeland and were held in the greatest esteem and honour by their peers because of the miracles they provided. And it also happened that some of them (upon whom people bestowed divine honours and worship, and for whom they set up graven images and statues, namely Apollo and Aesculapius) were regarded as gods by the pagan world. Similarly, Euperices and Herosilus were regarded by their peers as demi-gods. Thus also Hippocrates and Chrysippus acquired from the Greeks the highest honours as sanctified ones and saviours of the body. And, although this pagan superstition of those times must be regarded by us Christians as abominable, nevertheless we are never able to admire and praise highly enough those remarkable gifts and arcane mysteries of these distinguished physicians who were able to arouse such notable admiration even in the hearts and minds of the uncivilized and barbarian people.

Allow me to conclude the first part of the present Declaration with, as it were, one word, inasmuch as I am conscious of how precious each moment of time is for Your

Most Serene Majesty, having in your hand so many matters of great importance; and I fear lest a discourse, wearisome to you, be longer than this vile matter; but I say that many wondrous secrets lurk in the repository of nature, which not even to this day have been revealed to us. And since the honourable and lawful inquiry into the mysteries of nature has as yet by no age ever been prohibited or interdicted for the Philosopher and Physician, I, who profess to be a Philosopher and Physician (although not the most distinguished) have decide to offer this brief Declaration of mine to Your Most Venerable Majesty (even if I had not been urged to do so by your gracious suggestion) so that I may give satisfaction to Your Majesty concerning the reasoning of my previously published *Tractatus Apologeticus* and thus, after the good grace and favour of Your Majesty have been obtained, I may live among my Muses happily, cheerfully and free from envy, filled at last with unexpected joy.

In fact, concerning that which is important in my *Macrocosm*, I could write a very large volume (which I know would be wearisome for Your Majesty to read) in defence of each part of it. But among all other things here I will disclose the reason why I have dedicated my book to Your Majesty. I acknowledge that I dedicated this book especially to God my Creator, because it is through Him that we act on this earth, breathe, live and possess all things that we enjoy in this life. It is certainly thus that whatever good is begun or completed by us must be particularly due to Him. Then in the second place, I have dedicated this book to Your Majesty because immediately after God, I acknowledge that the zeal of men and all their efforts ought to proceed to the honour of Your Majesty. Consequently, I have considered that it was permitted for me to depend upon your regal grace and gentleness to such a degree that in return for so many late night studies or so much work, I expected above all the grace and favourable view of Your Majesty.

Moreover, since a letter sent by a certain friend I realized that there was some controversy between the individual to whom I entrusted this volume in England[2] and the engraver and printer[3] concerning the dedication of my work. While the former endeavoured to assign the honour of my book and labour to the Landgrave of Hesse, the latter individuals in fact endeavoured to assign it to the Count Palatine, their own prince, and (as I have here some witnesses to this matter), at last I was compelled to transmit, unexpectedly, that twofold dedication, namely to God and to you, my King, so that I might absolutely prohibit them from assigning these works of mine to any mortal except to my King alone, to whom I acknowledge I owe what is mine. And this is the reason that I, a fosterling, have chosen you alone before any foreign prince as my Mecaenas and patron, not by any presumption, but induced by a love of Your Majesty, who is far better deserving than those who would have been substituted for you. But so that Your Majesty may realize that I have not done this rashly or indiscreetly, it will be well for Your Majesty to know that I received several letters from Germany informing me that men of letters, particularly of that country, and the learned of every profession, both Papist and Lutheran as well as Calvinist, praised far beyond my merits this volume of mine and seem to approve of my works unanimously. Whence considering that no one in the entire world was more worthy of the honour and dedication of these labours of mine, which have gained so much approval from the learned, than Your Majesty, my own King, who is quite distinguished in both sciences and letters; to him especially, before all others, I have willingly and with pleasure dedicated these works of mine. And although it may seem foul and indecent to praise oneself and commend ones own works, nevertheless, I humbly entreat Your Majesty not to impute to me the charge of vainglory if by the testimony of foreigners I should protect my reputation, which is at stake in this work of mine — the *Macrocosm* — before Your Majesty, and should defend the sincerity of my writings. Therefore, for the sake of brevity, I thought it worthy to place before the eyes of Your Majesty only the relevant portions of some of the letters sent to me from foreign countries lest the testimony of those previously mentioned by me seem to be made up or worthless. Therefore, first of all in supplication I do entreat Your Majesty not to reject the information that the engraver-printer, before he was willing to undertake this *Macrocosm* of mine, was eager to know the opinions of many men of letters, both Papists and Lutherans as well as Calvinists, and to find out their feelings concerning this volume. Accordingly, it is possible to see this from a portion of this letter:

> Concerning your great volume, before the printer was willing to undertake it, he showed it to many other learned men, who indeed greatly praised your work: furthermore, he even showed it to the Jesuits who come here in great numbers at [book] fair time, and they added their own stamp of approval, with one exception: they felt the work was more worthy of publication if the topic of Geomancy was deleted, which they, as you know, condemn according to their religion. But we do not consider this judgement of theirs to be of the least concern, etc.
>
> 20th April:
In the year of Our Lord 1617
Your most devoted,
Justus Helt[4]

In the same year I received through the hands of a certain foreigner completely unknown to me another letter written by a certain most learned man; I have put forward here to your Royal eyes a portion which pertains to my writings:

> By exhibiting to all men a beautiful example of your genius, learning and diligence, you certainly do what is befitting a good man and a philosopher. And certainly it is most reasonable that all good men should hold in high esteem, even the ignoble who are adorned with every type of erudition, but particularly those who are distinguished by modest of mind, temperance, the true religion and piety. There is not so much which I could write about you; for it is likely that you already know it well. In distinguishing the harmony of mundane music you profess to be

the most practised and experienced; and this is perhaps not undeservedly, etc.

December 19th 1617
Most dedicated to you,
Du. Bourdalone[5]

Additionally, I have received a note written by a certain distinguished doctor from Germany, a part of which I have inserted here:

Your writings on the arcane philosophy are very much approved of by their most profound followers. Your works therefore will also constitute additional things which pertain to the Macrocosm and Microcosm, and lead to the very centre of things. For undoubtedly you will give an opportunity to other more occult philosophers to come forward into the public forum, so that at least when these strengths are thus joined together, in this Saturnine age, it will be possible for us, the inquirers, to investigate, with open eyes, the heaven and earth as well as all of Nature disrobed of her garment, etc.

Done in Anhalt
31st December 1617
Your most observant
Matthias Engelhart
Philosopher
and Doctor of Medicine
at Anhalt[6]

From these it is possible for Your Majesty to perceive that even my Physical history has been very well received and accepted by investigators of Chemical Philosophy. This most distinguished Doctor gives excellent testimony to this matter. Also, for the same purpose, I received another letter from a certain Doctor of Law, in French, this small portion of which I, your most loyal subject, have selected for Your Majesty:

Please measure my affection by the effects of my devotion, which being aroused primarily by the admiration for your noble mind, advanced not only by the continual consideration of that which is in your treatises and works; the richest part of which, in the manner of the rising sun, now casts its rays on our Germany, as its light does through all the places of the most learned of Europe.

Your most obliged friend and servant

From Vienna 3 February   Jean Balthasar
in the year 1618   Ursin Bayerius[7]

And although I could offer Your Majesty many similar testimonies selected from the letters of others who wrote much good about me and my works, for the sake of brevity all others have been omitted; only by this note from a most distinguished German and Professor Primarius of the new University of Giessen in Germany shall I impose an end to this Declaration of mine to Your Most Serene Majesty and to the defence of my work, the *Macrocosm*. His letter is as follows:

Relying on the common bond of the study of Philosophy and Medicine, by which we are linked although otherwise separated by great distances, I have ventured to write this letter to you, most noble and excellent sir, patron much worthy of honour, particularly since an opportunity has been given by the present candidates in medicine, who until now have been my house guests, etc., and have departed from our University in order to pay their respects to your England and particularly to you. Moreover, as your noble work, the *MACROCOSM* (which you have published with the immortalizing of your name), sufficiently and even more gives evidence of singular benevolence and eagerness to help all men of letters to such a degree that I doubt not that you are also quite willing to favour the present learned young men.[8] According to Cicero, it is a virtue to love that which is invisible, even as now happens to you among all good men. For who does not commend your singular skill in the investigation of things,[9] when you recall from Hell, as it were, the true principles of things, which Hippocrates of former days in some places treated unclearly, and Paracelsus of recent days dealt with superficially; furthermore, you use a restraint in refuting the opinions of others which is rare amongst us. Only continue, worthy sir, do not suppress the description of the Microcosm, whose title promises not vulgar things, complete the eternal monument to your name such that it may be accomplished with felicity; and I pray that you live for a very long time and with good fortune. Written hurriedly at the new University of Giessen in Hesse.

10 August 1618

Your most devoted

Gregor Horstius

Doctor of Medicine

Professor Primarius and

Chief Physician to

Prince Ludwig of Hesse[10]

Certainly from the above letter from this most celebrated man it is sufficiently obvious that my opinions are not new, but rather are the most evident explications and most clear demonstrations of the secrets of nature which have been concealed or hidden by the ancient Philosophers under the guise of allegorical riddles and enigmas. On behalf of my self and this book of mine I could mention many other things filled with the most profound prudence, knowledge and judgement, which, lest I offer tedium to the ears and eyes of Your Majesty, I will omit, provided only that I am satisfied that the brilliance of the writing of these other learned men will show the measure of their approval for my works corresponding to their good will toward them, and will make clear to Your Majesty their opinions about my volume, the *Macrocosm*, by this sincere (as it were) Declaration.

Accordingly, O Most Serene King, I humbly entrust my cause to Your Majestys distinguished wisdom and singular kindness, as well as your judgement and just government which are filled and adorned with profound knowledge. Your Majesty, who, by asserting prominent evidence of your prudence and justice, certainly informs all men that, although Astraea (that chaste and virginal goddess of justice), in abandoning the iniquitous earth was borne away on sublime wings, has ascended into the star-bearing palace of the heavens, bearing with her the scales of justice; nevertheless, she, moved with pity towards us mortal Britons, has left behind her progeny, descendant, or as it were, her own Imperial offspring, who will always judge rightly concerning disputed political matters, and will govern with kindness; and in fact in whom the conspicuous gifts of wisdom are so firmly rooted, and so deeply implanted, that, just as he himself blazes with the royal desire to exercise justice, he is provided with Natures instinct for giving judgements and is thus well prepared in heart to settle whatever is in doubt, inasmuch as his learning and knowledge has rendered him capable of anything.

O Most Serene King, endowed with an ineffable grace and majesty, the unexpected splendour of your grace and serene appearance of Your Majesty has moved me, your obedient servant and most unworthy subject, and has excited immense joy in me; therefore I humbly entreat that Your Majesty, with the same grace and great richness of mind, not desist from accomplishing that work most full of divine love, and while I, turning all efforts of my life and studies to the praise of God and honour of Your Majesty, not cease to pour forth prayers to the Creator on behalf of your safety, the security of the Kingdom, and the success of every undertaking. And indeed, may there be praise, honour and glory for ever for my King and my Creator, who is both eternal and necessary to support life.

END

Notes

1. The identity of Justus Helt has not been determined.
2. Quite possibly Michael Maier. See Introduction.
3. Johann-Theodor de Bry and Hieronymus Gallerus in Oppenheim. See Introduction.
4. This portion of Helt's letter was also included in 'A Philosophical Key', fols. 12^v-13, with Fludd's English translation as follows:

As touching your great Volume, before the printer would undertake it, he shewed it unto many other learned men, which did very much commend your work; Also he made the Jesuits acquainted with it, who in numbers resort unto the fayer of Frakford [Frankfurt], which adding also their spur to your commendations sayed, that, only on[e] thing excepted, it was a work most worthy of edition, namely if Geomancy were omitted; the which science (as you know very well) they mistake of for their religions sake: But we esteeme not of this their latter judgement.

5. The Lord of Bourdalone was the Chief Secretary to Charles of Lorraine, Fourth Duke of Guise. Fludd styles Bourdalone as *amicus meus*, and it was probably the noble secretary who arranged for Fludd to tutor the Duke and his brother (see Introduction). Fludd, *Anatomiae Amphitheatrum*, p.233, English translation in John Webster, *A Displaying of Supposed Witchcraft* (Jonas Moore, London, 1677), pp.319-20.

6. Matthias Engelhart was the town physician of Aschersleben, located between Magdeburg and Halle in Saxony. Aschersleben was the town built adjacent to the ancient seat of the Princes of Anhalt. Fludd included this letter in 'A Philosphical Key', fols. 19-19^v, with this translation:

> Your writings ar very wel approved by the most inward and affectioned searchers of the secret and mystical Philosophy. It wil therefore be your part fully to publish the rest, that belongeth as wel to the great as little world; for in so doing you wil shew other secret and concealed Philosophers the way to appear openly and show themselves to the world, that thereby all their forces being so united, it may at last be permitted unto us (the diligent searchers after Nature) even in this Saturnine age to knowe heaven and earth, and so to uncase and discover that universale nature, which is masked in darkness.

7. The identity of the writer of this letter has not been determined.

8. In his translation in 'A Philosophical Key', Fludd changes this line to read 'thes two learned young men'. See note 10.

9. Here Fludd's translation reads 'inquisition of naturall mysterys'. See below.

10. Gregor H. Horst (1578-1636) was undoubtedly the most esteemed and famous of those whose letters Fludd incorporated in the 'Declaratio'. He received his MD from Basel in 1606, and

became a Professor of Medicine at Wittenberg. In 1608, the Landgrave of Hesse called him to the University of Giessen and made him his personal physician. Horst gave up his professorship in 1622 to become the town physician in Ulm, where he remained until his death. In his time, he enjoyed a great deal of fame. He published numerous books, mostly medical, and his contemporaries hailed him as the 'German Aesculapius'. See August Hirsch, *Bibliographisches Lexikon*, 2nd ed., 5 vols. (Urban and Schwarzenberg, Berlin and Vienna, 1929-34); Vol. 3, p.304. Fludd also included Horst's letter in 'A Philosophical Key' (fols. 10-11), which reads this way in his translation:

Worthy Sir and my much respected friend, (relying on the common bond of study in Philosophy and Phisick, by the which we (being otherwise desioyned by a great distance of place) are united together), I thought it fit at this time to wright unto you especially occasion being offered by thes present candidates in Physick, which have been heatherto my household guests, etc. who departed from our University on set purpose to see England, and principally to salute you. And as your notable work of the Great World (which you have published with a perpetual eternishing of your name) hath expressed abundantly your singular good will and love towards such as are learned: so I doubt not, but that you will readily favour and assist thes two learned young men. Cicero sayeth, that it is the act of vertue to love thos things we see not, even as now it happeneth to you amongst all good men. For who doth not commend your singular industry in the search and inquisition of naturall mysterys, seeing, that you have revoked as it were out of the pit of hel thos true principles, which Hippocrates long since in some places of his works, and Paracelsus of late dayes have handled but superficially; using that modesty in refuting the opinions of others, which is rare and but seldome found among other men. Go forward then, worthy sir, and suppress not the description of the little world, whose title promiseth things, that are not vulgar, and by that means finish the eternal monument of your renowne which that you may accomplish with felicity — live happily.

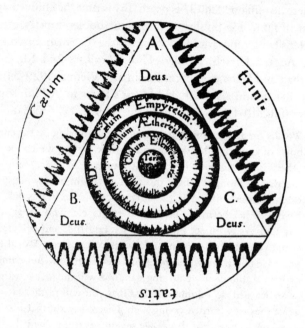

Figure 1: The Divine Triangle.

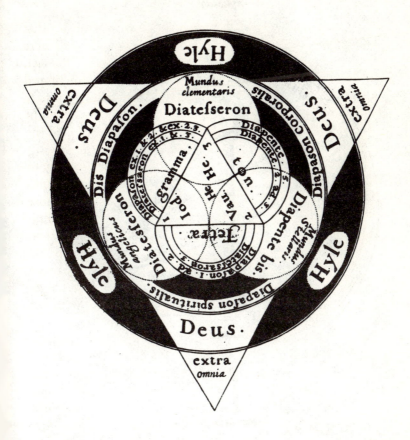

Figure 2: The Emblematic Manifestation of the Trinitarian Nature of the Universe.

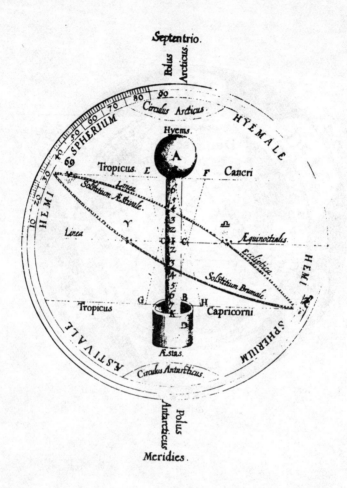

Figure 3: The Weather-Glass shown in Relation to the World.

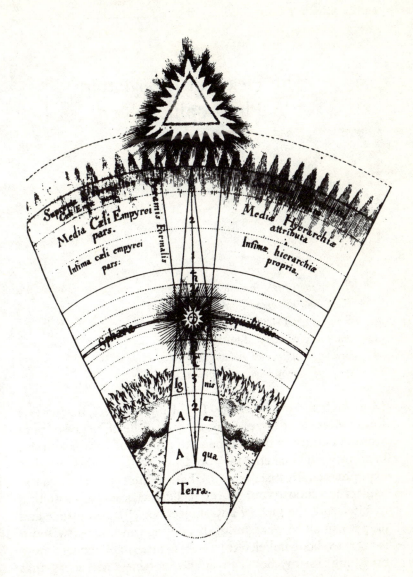

Figure 4: The Three Heavens and the Interpenetrating Pyramids of
Form and Matter.

The Heart of the Matter: 'A Philosophical Key'

*To the most high and mighty
King, and my dread Lord and
Sovereign, JAMES, by the
grace of God King of great
Britain, France, and Ireland etc.,
long life, happy reign, and eternal prosperity.*

Most high and mighty King, I your poor subject, having ever found your most sacred Majesty out of the affluence of your wonted bounty to behold me with the eye of your favourable benignity, it maketh me bold as well in this English experimental Treatise, as in my Latin Philosophical discourse to elect and choose your most Excellent Majesty (surmounting in Literature the pick of any subject, excelling in justice and judgement all Magistrates of this our age and whose wisdom is not to be paralleled by any Prince or Potentate now breathing) for my best Patron and worthiest Mecaenas, that under the shadow of your most royal Wing (as being sheltered and covered with a sure buckler and rock of defence) I may shield myself from the Harpys talents and envious endeavours of the world. Which also I have presumed to attempt so much the rather, because this containeth (as it were) the Key to unlock and open the meaning of that Macrocosmical and Microcosmical Philosophy, which before this time I have devoted unto you Majesty. Glad him therefore (I most humbly beseech you) with your wonted favour, and enrich him with your most gracious acceptance, who will for ever live and die

Your Ma^{tys} thrice humble and
most Loyal subject
Rob:Fludd

To The Reader

I am sure, that it is not unknown unto you (Courteous Reader) even from your infancy, how strangely and variously the dispositions of men are affected, seeing that some of them are feathered with ambition, others fixed with jealousy, many repining with envy, and most of all choked with dissimulation, and as it were, drowned and smothered in the vanity of self will, leaving but a very few to erect an eternal trophy or monument unto virtue, and to embrace Justice, tracing in the sincere steps and modest parts of Temperance . . .

And as for me, although I have felt the very selfsame blast of envy in many of mine actions, yet have I secretly smiled with my self, and rejoiced to think, that mine inward conscience . . . doth solace and comfort me, in justifying herself to be altogether ignorant of mens slanderous taxations and imaginary reports . . . Even thus have I breathed in silence . . . But perceiving Dame Charitys harvest to be more and more choked up with unprofitable darnel and fruitless taxes . . . I thought it at this time most expedient to break my wonted silence, and with my pen to satisfy mens doubts, which are quickly prone to believe, where no reason is produced to the contrary.

And now the green and fresh wounds of a certain silly and poor Apology of mine (written against D. Libavius, as well to set my unable wits on the tenterhooks to make trial what hue they would bear in the world, as also to insinuate into those learned and famous Theosophists and Philosophers acquaintance intitled the Fraternity of the Rosy Cross, in whose defence it was by me written) . . . Hence proceedeth it that out of the marrow of that little book these vain delusions and chimeras of suspicion do float amongst men . . . For truth . . . hath found out, that mine Apologys greatest accusers have been for the most part such, as have least pierced into the essence and secret of her subject, much less have they by due inquisition censured the meaning of her lines: for thus much some of them (being questioned) have ingenuously and voluntarily confessed, namely that they never read or saw her, but heard so much. Is not this (Just Reader) a true character of envy? Some of them again have read it, but not understanding the tenth part thereof (as by just inquiry it hath been plainly discovered) suspect my religion; alleging that the places of Scripture by me produced are falsely and erroneously cited and strangely wrested from their true sense, and thereupon have not stuck to blaze abroad, that my Philosophy is both spurious and corrupt, as also that my Theosophy is indirect. Is not this, I say, an infallible argument of malicious ignorance?

Another learned and well-understanding gentleman, who protesting that he had read my Apology over and over again did not stick to affirm in mine own hearing, that absolutely and without exception I had maintained the Fraternity of the Rosy Cross' assertions, until I had produced to his sight these evident caveats and exceptions expressed in that mine Apology, fol. 6: These men without all doubt (if they be free from devlish superstition etc; and fol. 9: All occasion of suspicion is to be exempted from them (if they speak in good earnest and as they think); and fols. 15, 16: Which being duly considered, we find many testimonies of the holy Spirit in these brethren (if their deeds do agree with their words); and many other places through the whole file of my book I produced, which easily satisfied him and made him to acknowledge his error.

Then said I to my self, (Good God) how slenderly is Charity and Love maintained amongst men, when discords of one mans words or writings are so spitefully perused by another, that the notes of concord and union, which should make all upright, are so blindly overslipped? Why should one image of God so rejoice at anothers error and misfortune, as maliciously to adulterate by a willful misprison that which is already in itself good and righteous? Judge therefore and witness in the first place ye heavens and that everliving power whose tabernacle ye are, if I ever knew or contracted my wishes unto any religion, saving this reformed one, so happily celebrated here in England!

. . . if I have written as well Theosophically as Philosophically in the defence of these renowned Philosophers, must I therefore seem an Innovator of Religion? Ought I consequently to know or be acquainted with them, because I write conditionally in their defence? Or is it folly in me to maintain truth and register with my pen mens virtues, whom I never knew, when by the learned precepts and sage advice of Cicero, we are taught to love those things, which are virtuous, though we never saw them?

Neither have my bigger volumes of each World (I mean as well Microcosmical as Macrocosmical) escaped the bitter assaults and sharp censuring darts of their uncharitable speeches, who have raised up against them for their better approaches unjust rampiers of offence, surnamed in the ears of the multitude innovation, new principles and opinions, spurious Philosophy, and strange doctrine never heard of before . . .

For in the first place to maintain and confirm mine opinions I cite and allege authority, as well sacred as Philosophical, and therefore doubtless they are not newly by me invented. Then in the second rank I produce infallible axioms and perspicuous reasons out of ancient and true Philosophy to approve and uphold mine assertions and thereby it is apparent that my doctrine is neither so irregular or unreasonable as they would make it. Thirdly, I have declared that the foundations of my Cosmical Philosophy is all one with those of the most high Prophet and prime Theosopher Moses, that is to say, Darkness, Water and Light; from whom Aristotle deriving craftily his Principles did first wrongfully attribute the invention of them unto him self, and then (the better to mask and hide his intent) did impose other names or titles upon them, as First Matter (under which privation is also comprehended), Second Matter, and Form. And lastly I conclude with that which is, as it were, a main impediment, or negative bar to hinder and stop the forward proceeding as well of the malicious calumniator as ignorant detractor, for as much as I both prove and make plain all the effects of my Philosophical doctrine by easy familiar and ocular demonstrations, the very pointing finger of experience, which is able to instruct the rude and rustical, yea, the very fools themselves.

And all this not only many unpartial surveyors of my labours here in England but also men singularly learned beyond the seas have (to my no small contentment and comfort) found so far distant from idle fiction and crooked irregularity, that it hath pleased them to attribute unto my labours a better applause, than myself did ever expect at their hands. Yea and some of them have not stuck with the faithful dent and point of a public and well-tuned pen to register the characters of truth; averring and expressing thereby their inward and secret thoughts in vulgar demonstrances and hieroglyphical types, to signify and manifest unto the world, that in their conceptions my Philosophical Works contain, as it were, a revelation or producing of those mysteries of Nature to light, which the ancient Philosophers have hidden from the worlds eye by masking and obscuring them with aenigmatical shadows, parabolical types, and ambiguous riddles, that thereby they might shelter and hide them from the knowledge as well of the foolish and ignorant, as of the unworthy person . . .

[Here Fludd inserts the testimonial letters from Gregor Horst and Justus Helt that are in the 'Declaratio Brevis'.]

By which lines I am sufficiently advertised that my volumes are neither distasteful to the Doctors of the Calvinists among whom the printer liveth, nor Lutherans, which are his bordering neighbours, no, nor the Papists, being that the

Jesuits themselves (who for their profundity in Philosophy have been especially respected over all Europa) have allowed and approved them. Neither have the truly literated persons of this our country (which have thoroughly perused my Cosmical history) smothered their good opinions concerning it, but have freely and ingenuously confessed and acknowledged what their judgement and censure is of it, and thereby have disannulled those ridiculous reports, which my invisible detractors . . . have buzzed abroad into the ears of many . . .

For ceasing not to aim at my writings with the arrows of their slanderous tongues, they, being induced by a far contrary affection, attempt now by overmuch attributing unto them, to derrogate from mine honour and reputation . . . Their later invention therefore (as I have been by many most credibly informed) hath been to divulge abroad that without all question I had the familiarity of some of the before specified Fraternity of the Rosy Cross: for (say they) we hold it impossible that he should entreat and write of such mystical learning (as they please to term it) and to be conversant in so many sciences, except other notable heads were joined with his in such a tedious business. Amongst the rest, this very concept was put into the head of an honourable personage and peer of this realm by one who seemeth a great scholar and doctor. But saving his credit and their honours, which are so prompt and ready in the deprivation of my little worth to spread abroad such fantasies, I do assure them, and with them the whole world besides, upon my faith and reputation, that hitherto I have not borrowed any mans assistance to erect any of my two Cosmical edifices, which seem so strange to those, whom either idleness will not suffer to peruse, or envy to acknowledge; but have soared, and do yet fly on mine own wings . . .

To confirm therefore by an evident declaration that my Philosophy hath flowed only from mine own invention as well in practice as speculation, I thought it (Gentle Reader) most convenient and necessary to express in this present Treatise, for the worlds better satisfaction, a certain ocular demonstration or experimental conclusion . . . that thereby every one may the better understand my meaning, ponder with ease the integrity of my heart, and at large . . . decipher the corrosive infirmity of my detractors and calumniators.

Moreover I do boldly assure the jealous world before His sacred presence, who made it, that my Macrocosmical history as well natural as artificial was composed by me some four or five years before the renown and fame of the Fraternity of the Rosy Cross had pierced mine ears, as by the testimonies of my worthy friends Mr Dr Andrews, and that most learned Gentleman of the Inner Temple Mr Selden it will easily be justified.

. . . if the said Fraternity of the Rosy Cross be so excellently informed in the mysteries of God and Nature as Europa declares them to be, my works are as far from wonder or desert in respect of theirs (being only builded on the sandy ground of opinion and outward experience) as heaven is distant from the earth, or light from darkness.

Professing also unto every one before the everliving God (who knoweth each mans secret imaginations) that to this hour I neither have unto my knowledge ever seen, known or conferred with any of the said Fraternity so admired abroad of many, neither understand I of their orders or conditions more than I have gathered out of such public books as I have read. And thus much do I utter unto my grief, openly acknowledging that I have earnestly desired their society and familiarity for the better erecting and exalting of my mind in the knowledge of mysteries as well divine and natural; but finding now by so long trial that mine Apology made in their defence hath no way moved their affections towards me, neither hath had that magnetical force, as to allure them unto mine acquaintance and conference in all this space, I do firmly imagine that they perceive some notable unworthiness in me to participate in their high mysteries, and therefore do, as it were, now despair to attain unto that, which so many thousands besides myself have so privately and publicly desired, and in conclusion failed with me in their wishes (as by daily discourses and books it may well appear).

This (I say) is the very cause that I content myself with the speculation of mine own experimental conclusion, making it serve in lieu of an outward looking glass of Nature, in which I have a long time contemplated those principles and mysteries upon which the ancient and wise Theosophists and Philosophers have grounded and planted the foundation of that Pansophia or Universal Nature . . . For although in my Macro and Microcosm I have demonstrated the reasons of each world by many spiritual and chemical conclusions, yet I must confess, that this subject, which I handle in this present discourse, hath been the main practical perspical or looking glass, in which I beheld the most assured and continuated progress in Nature by which I composed the whole fabric of my Philosophy . . .

And although the exiguous capacity and great imbecillity of my wit hath adjudged me before the Fraternity of the Rosy Cross unworthy to ascend the hauty Parnassus of their exalted knowledge, and hath debarred me from the happiness to taste of the sweet and fiery nectar, which floweth from the bright streaming fountain of Helicon, trickling and springing first from the sweaty limbs of flying Pegasus, and from antiquity consecrated to the Muses; yet my sickly and despairfull affection is sufficiently refreshed with the hopeful balm of consolation which assured my understanding that their Pansophia or Universal knowledge in Nature is not in effect much dissonant or disagreeing from that speculation which I have gathered out of this my practice in Alchemy.

And this have I not only observed out of books by them published concerning that subject, but also by the express assertion and confirmation of a letter, which I received from a learned Doctor of Germany, who by his writings seemeth to be acquainted with the very same Cabalistical and secret Pansophy so much bruited abroad by the fame of the foresaid Fraternity . . .

[Fludd here inserts the same portion of a letter from Matthias Engelhart, a doctor of medicine at Anhalt, that he includes in the 'Declaratio Brevis'.]

By which lines it may well appear that this abstruse Philosopher deemeth me to participate with him in knowledge of the highest mystery of Nature, for as much as he acknowledgeth, that my writing doth bear shadow of truth. And verily I must confess that this mine experimental conclusion hath so cleared the masking darkness of mine accustomed ignorance, that I do . . . discern in the glimmering light or dawning of the day . . . that such a mystery and divine thing there is contained within the secret cabinet of Nature and her four elements, of which both the divine Theosophists and profound Philosophers have so often mentioned in their aenigmatical discourses and parabolical labours . . .

A Nosce te ipsum: To the Malicious Detractor or, the Calumniators Vision

[This is an allegorical tale, told in first person and in a flourishing style, of Fludd's 'vision' of Creation, how the macrocosm and microcosm developed, with the latter an exact reflection of the former, and how they are intertwined. This selection is from the conclusion of the section, where he admonishes his detractors to remember their divine origins and indeed, to continually go inward for guidance in acting in the outer world.]

Art thou (I say) the perfect example and pattern of that infinite Nature, and is it possible thou shouldest be so false and treacherous unto thy self? Wilt thou be still blind, and not consider, that each mans spirit (though included in diverse bodies) is made all one by the immediate, and first emanation or Universal beam of Unity; and that Union in Nature can no more be severed or disjoined, than the bright stream which issueth from the centre of the Sun; which nevertheless searcheth and visiteth every place and region of the horizon as well in heaven as in earth? Is it possible, thou shouldest suffer thine unnatural memory to be so great an enemy to thy souls repose, as to forget and blot out of his margin the divine and celestial race, from whence thy soul is sprung? Why doest thou permit thy self to be so fondly blinded by the prestigious delusions of Litigium [the first-born son of Chaos, i.e. the Prince of Darkness] . . . and rent into factions that precious unity of Soul and Spirit, which our Creator hath so linked together by the gentle and charitable tie or knot of amity?

. . . henceforth therefore it plainly appeareth, that all men are but Unity in essential virtue and act, the which is so united to the first actor in Nature, that it is altogether in truth inseparable and indivisible from it, although men and all things else divide and branch themselves into different persons and shapes in respect of their Macrocosmical materials; for number, distinction, difference and multitude are the defective progression of blindfold and deceitful matter . . . But in the subtle spirit and form . . . is stability, permanency and Unity, from whence only the sweet consonance of peace and the worlds harmony, is extracted . . .

Then how far wandrest thou (O fond Man) in the glimmering of darkness and error, which groundest thy faith on things subject unto the outward sense, which are only compositions of antipathy and discord, and neglectest to behold with spiritual eyes the central and hidden truth, where Unity abideth compassed about with gleeful joy and peaceful sympathy?

Return, return, (I say) unto thy self, subject thy body unto thy reasonable soul by diving into thy inward treasure, and then prostrate and submit . . . thy mental and spiritual part unto thy God; for so shalt thou be made one spirit with him, conditionally that thou doest persevere in humility, and acknowledge from thy heart that grace of thy Creator, by which thou shalt be glorified and exalted . . .

Wherefore I cordially admonish thee to ascend from this world unto God, that is, to penetrate quite through thyself, for to climb up unto God is to enter into thy self, and not only inwardly to visit thy dearest Soul, but also to pierce into the very centre thereof, to view and behold there thy Creator; which . . . thou mayest the better . . . gather and recollect the beams of thy inward man from the distractions of this outward world, and revoke them from external actions, to participate with thy inward joys . . .

This is the life of the spiritual man, firmly to contemplate God and to be sweetly delighted and refreshed in that . . . speculation; yea, verily it will be a most dear and pleasant apprehension to meditate on Him, love Him, and with admiration to adore and extoll Him in heart. Wherefore nothing can be more notable and excellent unto a blessed life, than by closing up the carnal senses . . . to convert every outward affection into the inward self, and to refer each foreign and alienated inclination and intention from the lustful appetites of mortal men unto oneself, and by that means only to confer, as it were, and talk with Demogorgon [God, the Creator], who dwelleth within *thee*.

Thus I say, wilt thou know, that each man is thy brother, and that thy brother is part of thy self, and all men are but one and the selfsame thing in specie, which is in effect the very unity . . . and essence of God Himself masked from the sight of unworthy men with the material mantle of Nature.

Of the Excellency of Wheat

The first book of wheat & bread in general, & of the excellency thereof

As the Universal Nature, sole mother of this great orb and every creature therein, is in herself pure, simple, transparent & invisible, so hath she in this lower region of the world three especial & notable Kingdoms compounded of her simple Elements to govern & inhabit, in the which she hideth, as with a mantle of obscurity, her secrets from the eyes of the unworthy. Every one of these . . . is distinguished from each other by . . . their Empress dwelling . . . in the centre or middle of them, where she guideth with equity and ruleth by due proportion every member of their circumference.. [She] inhabit[s] the Sun . . . & . . . send[s] out her glittering virtue . . . downward to the lowest earth as upward to the highest heavens so that all things might be equally . . . vivified by . . . her beams. In like manner also ruleth she . . . the heart in the lesser world [i.e. in man], distributing her vital spirits circularly, as from a virtuous centre to the circumference thereof by the streaming vessels of arterial channels. In the selfsame manner I say doth [rule] this most peerless Queen . . . this princely virgin & handmaid of the everliving Creator, the invisible fire of Zoroastes & Heraclitus, the essential Ligament of the Elements . . . which causeth so compleat an union among the four dissonant natures of every one of the foresaid Kingdoms in the sublunary world.

Think therefore seriously that she hath chosen for her chiefest mansion in the Animal Kingdom the body of Man, that most excellent of all sensible creatures; in her Vegetable Empire she hath elected Wheat that most worthy of all vegetables . . . & in her Mineral nature she most delighteth in & principally inhabiteth her golden palace [i.e. gold], burnishing it about with the steams of her brightest glory.

How excellent would that Artist appear beyond his companions, who by the most complete act & fire of celestial Alchemy, could first learn justly to distinguish the parts as well spiritual and corporal of the Queens chiefest . . . dwelling-place; for so by the surveying of his most secret and hidden regions he shall quickly be taught in a true vision to know himself and to discern the highest heaven of his inward Man . . . Also what happiness and joy shall be infused into the faithful operatour, whose good fortune it will be truly to . . . behold and perceive all the secret mysteries of Nature and her ministers . . .

And verily although I profess my self to be ignorant of that excellent skill either to understand rightly those hidden mysteries or the regiment of that aetherial fire by which such glorious effects in Nature may be accomplished, & therefore am debarred and denied from entering into that straight path . . . although (I say) the sharp punishment of Prometheus hath added terror unto my thoughts & deprived me of the hardiness to steal any of this excellent fire from heaven; neither hath mine ignorance & unworthiness permitted me to obtain it from above through grace . . .

. . . yet with the common & spurious Alchemist, that toyish ape, and superfluous imitator of Nature, I will here presume to dive so far into the hidden parts . . . of this rich and unvalued vegetable . . . being contented with the lot and parcel of that curious experience which mine own labours & industrious search into the secret natures of things have taught me, in which I (for defect of an higher agent) have elected the earthly fire & lower elementary heat to supply . . .

. . . [wheat] is observed by the most learned in physic to conform . . . itself by . . . a likeness . . . in disposition unto mans vital and natural faculty, and therefore nourisheth more those parts in him than any earthly vegetable . . . Each thing in and through his . . . included spark multiplieth infinitely; this is the means by which every vegetable, in itself corruptible through multiplication, is perpetuated. And verily I cannot find out in all the Vegetable Kingdom a more exquisite & excellent member for multiplication than the grain of wheat . . .

And hereupon it appeareth that wheat in his manifest quality is moist by reason of the abundance of humidity which it hath, and hot in respect of the abundance of natural heat . . . Next shall we perceive that in this our subject a limpid and subtile aery element lurketh, which containeth in it many abstruse and hidden faculties as in his due place I will manifest at large. After which the faeculent and stinking element of fire (the perfect agent to all corruption) appeareth . . . And lastly that glorious and aethereal spirit . . . will show itself leaving his dark prison . . . We shall therefore observe and find after a due inquisition these elements to be resident in bread or wheat, to wit: earth, which was the last faeces in the separation; then the water, which appeareth first; next the air which keepeth the second rank; and afterward the fire which followeth the air, whom the quintessence or fifth element followeth . . .

Neither would I in this place inform you that these are simple elements in their nature, for so it pleased his most excellent and learned Majesty to object unto me and that most rightly . . . but as I answered him so I tell you that these are called elements for as much as they participate for the most part of this or that element, for nothing almost in the whole course of Nature is found pure and neat . . .

As we have before expressed with the consent and authority of the most learned Philosophers that the superfluity of water in this vegetable hath proved the instrument of many diseases and inconveniences in mans body . . . so also is there found in it a perfect means of mans health . . .

What causeth bad humours to be so irregular and violent in mans body as to procure diseases in it, and by that means to hasten on untimely death if it be not prevented, but a want of that universal tie and knot in Nature whereby the four disagreeing elements are peaceably united in the outward mass [i.e. the fifth element or quintessence]; for as we see that one unruly horse in a stable being broke loose never ceaseth to jar and fight with the rest of his companions till he remain sole victor in the place, even so one element feeling the want of his quintessential bridle or ligament, breaketh forth from his rank, and never leaveth to persecute his contrary in the same composition till all the economical habit of mans palace be destroyed and the lifes spirit set at liberty; and all this we may perceive to proceed from the defect of those heavenly and sympathetical bands by which harmony and concord should be preserved in the whole bodily frame.

And verily there is a most assured token of death where the universal ligament of Nature seemeth to be violated, but if in part his quintessential knot be untied it signifieth mortification or deprivation of life in that same very part; and if there chance but a weakness or debility of it in one or more portions of the body, it then breedeth a malady or sickness by so much the more dangerous by how much their harmonical bands are attenuated or violated.

Which being so, it appeareth (as I imagine) most evidently that by nourishing and refreshing or those fainting or debilitated ties of Nature by their like, their impairing vigours may easily be repaired and rectified again . . .

And verily I must really confess that I once finding my self very much molested with an ache and debility in the back of mine own hand, and being at that very time occupied about this very same experiment which I handle at this present, I took a little of that crude quintessential balm of wheat and therewith did anoint the back of my hand, & from that hour being three years since till this very season, I never felt that pain wherewith I was ordinarily affected before that time. And yet I profess (with shame be it spoken) either my negligence or other occasions have been such that neither before nor since did I ever make any farther trial of the property thereof; for as much as in my spirit I have wished that by those experimented Philosophers who have been better practised and acquainted with the secret virtues and Nature of this fifth element, I might be fully & truly taught a way to exalt it to his uttermost effect of purity.

What this quintessential balm is . . . is the very spirit of the starry heaven replenished with the bright beams of the Sun.

[Fludd then speaks of the mystical significance of wheat in Holy Scripture, and continues . . .]

Of two kinds of bread whereof the one is terrestrial and vulgar, the other celestial & proceeding immediately from God [Manna from heaven]

What then in respect of this wisdom is that of man, but mere folly and self-willed ignorance by which no other true effect but deluding imaginations are effected. Blessed is the man therefore . . . yea that thrice blessed that possesseth and rightly governeth this bright heavenly Manna, well may be numbered amongst the elected sons of God who is thus beautified with the presence & virtue of Christ, for they enjoy all the treasure, both of health, wealth, and prosperity, as well spiritual as corporal . . . Neither do I doubt but that there remain such in the world, because the Holy Scriptures teach us that sapience hath her chief delight amongst the children of men . . .

And verily if the Fraternity of the Rosy Cross in whose behalf mine Apology was written, for the which I have been without due reason or occasion suspected and brought in some question, if they (I say) have been so fortunate through grace to have lighted upon this lovely creature . . . as they themselves do with the sound as it were of a trumpet profess unto all the world, I make no doubt but all ye that shall perfectly understand so much will assert with me that these are to be numbered among the rank of those of which Solomon speaketh . . .

He which attaineth wisdom is blessed; for the purchase thereof is better than the merchandise of silver or the revenues of gold . . . But if their reports be but vanity and wind . . . yet am I not wrong if I think that such a divine essence there is which communicates herself with some men, when I find myself to be maintained and . . . propped up in my opinion by the Holy Bible. Neither (since I know them not) did I ever aver constantly that they were so divinely possessed, or could do that they published, but did affirm in mine Apology that what they professed was possible to be accomplished, as well by natural as supernatural means.

To conclude . . . I say that this heavenly bread, this celestial Manna, this spiritual rock, this divine sapience, this all in all of life, Christ Jesus, is the second kind of bread which differeth from the common bread as heaven doth from earth, or light from darkness . . .

The Second Book Containing the Mysteries of Wheat or Bread.

[In this book, Fludd details his experiment with running commentary and macrocosmical-microcosmical connections. The procedure is to achieve putrefaction to break the wheat down into its constituent elements, then to observe the different qualities as he 'resurrects' them in generation. To do this he will imitate the natural processes of nature by using, as in the furnace of the earth, four different degrees of fire:]

. . . that is by a hyemal winter fire, with which she rotteth; & then with a vernal or spring fire by which it groweth & indueth his green mantle; & then with his aestival or summer heat, by which he flourisheth & putteth on a crest or helmet of a more golden colour, hastening by this intense agent unto his maturity; and lastly he changeth that fire to an autumnal or harvest operation by which the mark of maturity so aimed at all the year is touched on the very head that then corruption may again begin . . .

[Beginning with a gentle fire that resembles autumn, he elected to 'violate by corruption the bands of those four elements which did preserve that vegetable form until in a manner I had deprived it quite of all shape, & reduced the whole body unto a muddy or slimy substance . . .'

Having achieved putrefaction, Fludd has to work with greater intensity of heat and in a different order than nature to utterly destroy the putrefied substance and thereby separate its elements. Thus the next step is to apply intense heat, 'much like that of the parching summer, but of a far ungentler disposition', which results in the wheat's reduction to 'the first matter or original principles from whence it sprung and as derived at the first'. This 'first matter', or *prima materia*, corresponds in the macrocosm with the original Hyle or uncreated void and darkness out of which God created the universe.

Through the application of intense heat, a misty cloudy rose 'ten cubits in height', condensed into droplets and then collected as a small pool of water. Fludd observed that it 'was everywhere animated and adorned with the bright tincture of a spiritual light'. This was analogous to Creation where the spirit of God moved upon the waters.

The next step was to free the aerial spirit by switching to three different, but gentler heats to imitate the 'temperate disposition of the vaporous spring' as in March when the sun enters Aries; the second was even more hot and moist as when the sun enters Taurus; and last, most hot and moist as when the sun enters Gemini. By the first he drove off the thinner water from the spirit and then had a golden liquor; the second gave a spotted residue; and the third resulted in a 'a pure white Crystalline spirit composed all (as I could guess) of a pure volatile salt'. This aerial spirit is the agent of all growth and multiplication in plants and animals, is always found in water, and is analogous to the universal spirit. Rain causes growth because it is rich in aerial spirit. Further heating of the crystals resulted in their conversion into 'a million sensible Atoms flying in the air'. (Fludd believes it 'probable that all things were made of Atoms as some Philosophers have guessed'.) Fludd further concludes that 'the nature of this spirit is to purify'.

The next element, fire, was subsequently produced:]

. . . fire began to show itself in shape of a fiery tincted oil, quite different in nature from the water, and of another shape than the air; it was accompanied with a great stink and evil odour, & after his purification done by circulation or rectification, he seemed of a high saffron or ruddy colour, not unlike the red flame of a common terrestrial fire . . . I found therefore by due contemplation of this part, that it was that sole active element in this lower sublunary sphere which especially produced such strange alterations to corruption in the world, and that creatures were brought to putrefaction by it; & that when it reigneth most in the body to be putrefied, then is the same body most rotten . . .

[Next comes the fifth element, the quintessence.] Now are we ascended above the elementary heaven & entered into the region which the Alchemists call Quintessence, the Philosopher Æther or *materia simplicissima*, the Poets *ceolum*, *olympus* or *campus elyseus* &c. [This element is] the very life by which the four elements do act in mans body by virtue of the central or inward motive principle, through which the whole composition subsisteth . . .

[It was a difficult element to extract, for which he was forced] to invent a peculiar heat which gave at one side the vessel, by which I recovered this æthereal humour & purified it to his greatest brightness.

Then did I evidently perceive it to be of a strange disposition, for which the least heat it would be liquid & as red as blood, filling the whole vessel full of Atoms; and immediately by the coldness of the air be congealed into a perfect diaphan and transparent substance, like a precious stone, mingled partly with a clear crystal substance, & partly of the golden colour of the Jacinth, or as a crystal glass all to be seamed with sparks and streaming stars of light . . .

[Fludd here makes a distinction between soul and spirit.] The vital soul is a bright light from God which is carried as in a chariot in a clear spirit on whose wings it flyeth . . . the spirit is the female . . . and the light is the male or agent by which all things do live; and by that reason we [affirm] that the thin limpid and diaphan matter like white crystal is the vehicle of the soul . . .

[This quintessence is the substance Fludd put on the back of his hand to relieve the ache.]

[Last of all, the experiment is left with the fourth element, earth.] We find after the separation of the quintessence a dark mass, a *caput mortum* or a *terra damnata* to mans sight but not in effect . . . in this our earth we found a central heat remaining which is the very act & primary seed of every plant, mineral & vegetable, namely a fix[er] salt . . . All this vivifying but fixer salt I found in my earthy mass, and this fixer salt, according to his mediocrity or firmness in fixation, produced either vegetable or mineral . . .

[By speculation of his own and the writing of Hermes, St Peter and Moses, Fludd concludes:]

. . . and therefore we may conclude by consequence that earth is gross water, & water is gross air, & air again is nothing else but dense and thick fire. This therefore being proved by ocular experience to be true & certain it followeth, vice versa, that fire by condensation is made air, & air is thickened by the same operation into water & lastly water is coagulated into earth. If these therefore be so manifest unto mans sense then is it also evident that all things are but one thing, because all things are made of the four elements, which are in effect but that only one thing, which St Peter tells us to be water; and Moses doth not say that the spirit of the Lord was carried upon the heaven & earth, but upon the waters, by the separation whereof the heavens & the earth was afterward created.

We conclude therefore that the substance of the highest heavens is nothing else in a manner but a most subtle exhalation which consisteth of the thinnest water; & the middle heavens are as an exhalation of a middle composition, which is as it were a thin vapour . . . & the lower heavens are as a grosser vapour, which the higher it tends the thinner it appears; & the faecal substance of all this mass of water is earth. Whereby we may observe that all in this world is but water gross or subtle, & that all water is but spirit thick or thin, & therefore all things in matter is but one thing, namely water . . .

Cold is the principal actor in the condensation & coagulating of diverse disagreeing substances together, so also is heat the withstander and resister of those frozen actions of cold, whereby their bands are violated, melted & at the last quite dissolved into the mobile nature of gliding water. Hence therefore it proceedeth that where cold (the offspring of the dark Chaos and matter, & therefore an enemy and contrary unto heat, the issue of light and Demogorgon) hath greatest dominion, there reigns density & thickness in the general mass of the water; but where heat, [and] the mighty harbinger of the universal Emperor, light, is most present & potent, there again the contrary effect unto that of coldness is produced; where they meet & are equal in power, there is that sphere of equality, or orb and circle of the worlds soul, of which the Platonists have spoken so obscurely through which (as we have related in our *Macrocosm*) the glorious Sun of the great world moveth; or in which (as said in my *Microcosm*) the vivifying heart rideth in the little world . . . [See Figure 5, p.171.]

. . . We with other Philosophers [have] divided [the distance from the dark centre of the earth to the highest heaven] into six evidently differing natures, ascending (as by the scale of Jacob) from the centre of the great world up to the circumference thereof by these evident scales or degrees, to wit: Earth; Water; Air; Fire [all of which make up the Elementary or Lower Heaven or Region, the *Caelum Elementare*]; Æther [which makes up the Middle Region containing the sun, moon and planets, with the sun in the middle, the *Caelum Ætherium*]; *Spiritus Empyreus* [the Highest Region or *Caelum Empyreum*, which contains the angelic hierarchy]; and next above it we shall find out the seventh or highest one, where abideth the general sabbath of rest, where the presence of God is permanent ... [See Figures 6, 7 and 8, pp.172, 173 and 174-5.]

... this is the reason that I have compared in my Macrocosmical history heat and motion of the formal region of light & that of dark matter unto two several pyramidal figures whereof the cone of the material mounting upwards toucheth the base of the formal pyramid which is above, & the cone of the formal pyramid contrariwise descendeth downward, & passeth to the centre of the material pyramid bases ... [See Figure 4, p.99.]

[This experiment then became the basis for Fludd's entire system, with its many different applications, and elaborate detail.]

The Influence of Archetypal Ideas on the Scientific Theories of Kepler[1]

Wolfgang Pauli

6

Kepler's views on cosmic harmony, essentially based on quantitative, mathematically demonstrable premises, were incompatible with the point of view of an archaic-magical description of nature as represented by the chief work of a respected physician and Rosicrucian, Robert Fludd of Oxford: *Utriusque Cosmi Maioris scilicet et Minoris Metaphysica, Physica atque technica Historia* (first edition, Oppenheim, 1621). In an appendix to Book V of the *Harmonices mundi*, Kepler criticized this work of Fludd's very violently. Fludd, as the representative of traditional alchemy, published in his treatise *Demonstratio quaedam analytica* a detailed polemic directed against Kepler's appendix, whereupon the latter replied with an *Apologia* that was followed by a *Replicatio* from Fludd.

The intellectual 'counter-world' with which Kepler here clashed is an archaistic-magical description of nature culminating in a mystery of transmutation. It is the familiar alchemical process that by means of various chemical procedures releases from the *prima materia* the world-soul dormant in it and in so doing both redeems matter and transforms the adept. Fludd, unlike Kepler, had no original ideas of his own to proclaim; even his alchemical notions are formulated in a very primitive form. The universe is divided into four spheres, corresponding to the ancient doctrine of the four elements. The highest is the empyrean, the world of spirits, followed in descending order by the ether and the link with the sphere of the elements and sublunary things, and at the bottom, by the earth, which is also the seat of the devil. The world is the mirror image of the invisible Trinitarian God who reveals Himself in it.

The two polar fundamental principles of the universe are *form* as the light principle, coming from above, and *matter* as the dark principle, dwelling in the earth. All beings from angels to minerals are differentiated only according to their greater or lesser light content. A constant struggle goes on between these polar opposites: from below, the material pyramid grows upward from the earth like a tree, the matter becoming finer toward the top; at the same time the formal pyramid grows downward with its apex on the earth exactly mirroring the material pyramid [see Figure 4, p.99]. Fludd never distinguishes clearly between a real, material process and a symbolical representation. Because of the analogy of the microcosm to the macrocosm the chemical process is indeed at the same time a reflection of the whole universe. The two movements, the one downward and the other upward, are also termed sympathy and antipathy or, with reference to the Cabala, *voluntas Dei* and *noluntas Dei*. After the withdrawal of the formal light principle, matter remains behind as the dark principle, though it was latently present before as a part of God. In the middle, the sphere of the sun, where these opposing principles just counterbalance each other, there is engendered in the mystery of the chymic wedding the *infans solaris*, which is at the same time the liberated world-soul. This process is described in a series of pictures (*picturae*) which Fludd also designates as 'hieroglyphic figures' or 'ænigmata'. [See Figure 4, p.99.]

In agreement with old Pythagorean ideas, Fludd evolves from the proportions of the parts of these pyramids the cosmic music, in which the following simple musical intervals play the chief part.

Disdiapason = Double Octave	*Proportis quadrupla*	4:1
Diapason = Octave	*Proportia dupla*	2:1
Diapente = Fifth	*Proportio sesquialtera*	3:2
Diatessaron = Fourth	*Proportio sesquitertia*	4:3

This is expressed in the characteristic figure in Figure 6, p.172, representing the *monochordus mundanus*. It may be remarked that the idea of cosmic music also appears in the works of the alchemist Michael Maier.

Fludd's general standpoint is that true understanding of world harmony and thus also true astronomy is impossible without a knowledge of the alchemical or Rosicrucian mysteries. Whatever is produced without knowledge of these mysteries is an arbitrary, subjective fiction. According to Kepler, on the other hand, only that which is capable of quantitative, mathematical proof belongs to objective science; the rest is personal. It is already apparent from the concluding words of the appendix to Book V of the *Harmonices mundi* that Kepler had to fight in order to justify the adoption of strict mathematical methods of proof:

> From this brief discussion I think it is clear that, although a knowledge of the harmonious proportions is very necessary in order to understand the dense mysteries of the exceedingly profound philosophy that Robert [Fludd] teaches, nevertheless the latter, who has even studied my whole work, will, for the time being, remain no less far removed from those perplexing mysteries than these [the proportions] have receded [for him] from the accurate certainty of mathematical demonstrations.

Fludd's aversion to all quantitative mensuration is revealed in the following passages:

> What he [Kepler] has expressed in many words and long discussion, I have compressed into a few words and explained by means of hieroglyphic and exceedingly significant figures, not, to be sure, for the reason that I delight in pictures (as he says elsewhere) but because I (as one of whom he seems to hint further

below that he associates with alchemists and Hermetic philosophers) had resolved to bring together much in little and, in the fashion of the alchemists, to collect the extracted essence, to reject the sedimentary substance, and to pour what is good into its proper vessel; so that, the mystery of science having been revealed, that which is hidden may become manifest; and that the inner nature of the thing, after the outer vestments have been stripped off, may be enclosed, as a precious gem set in a gold ring, in a figure best suited to its nature — a figure, that is, in which its essence can be beheld by eye and mind as in a mirror and without many-worded circumlocution.

For it is for the vulgar mathematicians to concern themselves with quantitative shadow; the alchemists and Hermetic philosophers, however, comprehend the true core of the natural bodies.

By the select mathematicians who have been schooled in formal mathematics nature is measured and revealed in the nude; for the spurious and blundering ones, however, she remains invisible and hidden. The latter, that is, measure the shadows instead of the substance and nourish themselves on unstable opinions; whereas the former, rejecting the shadow, grasp the substance and are gladdened by the sight of truth.

But here lies hidden the whole difficulty, because he [Kepler] excogitates the exterior movements of the created thing [*res naturata*: the actually existing natural object] whereas I contemplate the internal and essential impulses that issue from nature herself; he has hold of the tail, I grasp the head; I perceive the first cause, he its effects. And even though his outermost movements may be (as he says) real, nevertheless he is stuck too fast in the filth and clay of the impossibility of his doctrine and, perplexed, is too firmly bound by hidden fetters to be able to free himself easily from those snares without damage to his honour and ransom himself from captivity cheaply. And he who digs a pit for others will unwittingly fall into it himself.

Such a rejection of everything quantitative in favour of the 'forma' (we should say for 'forma' symbol) is obviously completely incompatible with scientific thinking. Kepler replies to the above as follows:

> When I pronounce your enigmas — harmonies, I should say —
> obscure, I speak according to my judgement and understanding,
> and I have you yourself as an aid in this since you deny that your
> purpose is subject to mathematical demonstration without which
> I am like a blind man.

The disputants, then, can no longer even agree on what to call
light and what dark. Fludd's symbolical *picturae* and Kepler's
geometrical diagrams present an irreconcilable contradiction.
It is for example easy for Kepler to point out that the
dimensions of the planetary spheres assumed in Fludd's figure
of the *monochordus mundanus* [see Figure 6, p.172] do not
correspond to the true, empirical dimensions. When Fludd
retorts that the *sapientes* are not agreed as the ultimate
dimensions of the spheres and that these are not essentially
important, Kepler remarks very pertinently that the
quantitative proportions are essential where music is
concerned, especially in the case of the proportion 4:3, which
is characteristic of the interval of the fourth. Kepler naturally
objected, furthermore, to Fludd's assumption that the earth
and not the sun is the centre of the planetary spheres.

Fludd's deprecation of everything quantitative, which in
his opinion belongs, like all division and all multiplicity, to the
dark principle (matter, devil), resulted in a further essential
difference between Fludd's and Kepler's views concerning the
position of the soul in nature. The sensitivity of the soul to
proportions, so essential according to Kepler, is in Fludd's
opinion only the result of its entanglement in the (dark)
corporeal world, whereas its imaginative faculties, that
recognize unity, spring from its true nature originating in the
light principle (*forma*). While Kepler represents that modern
point of view that the soul is a part of nature, Fludd even
protests against the application of the concept 'part' to the
human soul, since the soul, being freed from the laws of the
physical world, that is in so far as it belongs to the light
principle, is inseparable from the *whole* world-soul.

Kepler is obliged to reject the 'formal mathematics' that Fludd opposes to 'vulgar' mathematics:

> If you know of another mathematics (besides that vulgar one from which all those hitherto celebrated as mathematicians have received their name), that is, a mathematics that is both natural and formal, I must confess that I have never tasted of it, unless we take refuge in the most general origin of the word [teaching, doctrine] and give up the quantities. Of that, you must know, I do not speak here. You, Robert, may keep for yourself its glory and that of the proofs to be found in it — and how accurate and how certain those are, that, I think, you will judge for yourself without me. *I reflect on the visible movements determinable by the senses themselves, you may consider the inner impulses* and endeavour to distinguish them according to grades. *I hold the tail* but I hold it in my hand; *you may grasp the head mentally*, though only, I fear, in your dreams. I am content with the effects, that is, the movements of the planets. If you shall have found in the very causes harmonies as limpid as are mine in the movements, then it will be proper for me to congratulate you on your gift of invention and myself on my gift of observation — that is, as soon as I shall be able to observe anything.

The situation is, however, not so simple as Kepler here represents it to be. His theoretical standpoint is, after all, not purely empirical but contains elements as essentially speculative as the notion that the physical world is the realization of pre-existent archetypal images. It is interesting that this speculative side of Kepler (here not avowed) is matched by a less obvious empirical tendency in Fludd. The latter tried in fact to support his speculative philosophy of the light and dark principle by means of scientific experiments with the so-called 'weather-glass'. Since this attempt casts light on what seems to us an extremely bizarre episode in the intellectual history of the seventeenth century, I should like to say something more about it at this point, although the relevant passages are only to be found in a late work of Fludd's, the *Philosophia Moysaica* (Gouda, 1637), which did not appear until after Kepler's death.[2] The weather-glass was constructed by immersing a glass vessel, opening downward, into a receptacle filled with water. The air contained in the vessel having been rarefied by heating, a column of water will rise within, its level determined by both temperature and air pressure. The latter concept, however, was not known before Torricelli, and the temporary variations in the water level, caused in part by variations in the air pressure, were usually interpreted as resulting from only the variations in temperature. On being warmed the column of water falls, on being cooled it rises, as a result of the expansion or contraction of the air remaining above the column of water. The instrument, a kind of combined thermometer and barometer, behaves, of course, in a way opposite of what we are used to.

. . . quotations from the *Philosophia Moysaica* make it apparent how Fludd regards the weather-glass as a symbol of the struggle between the light and dark principles in the macrocosm . . . [Fludd's description of the weather-glass is given in Chapter 8 on the *Mosaicall Philosophy*; see also Figure 3, p.98.]

It is significant for the psychological contrast between Kepler and Fludd that for Fludd the number four has a special symbolical character, which, as we have seen, is not true of Kepler . . .

From what has been said above the reader has gained, we hope, some understanding of the prevailing atmosphere of the first half of the seventeenth century when the then new, quantitative, scientifically mathematical way of thinking collided with the alchemical tradition expressed in qualitative, symbolical pictures . . .

Fludd's attitude, however, seems to us somewhat easier to understand when it is viewed in the perspective of a more general differentiation between two types of mind, a differentiation that can be traced throughout history, the one type considering the quantitative relations of the *parts* to be essential, the other the qualitative indivisibility of the *whole*. We already find this contrast, for example, in antiquity, in the two corresponding definitions of beauty: in the one it is the proper agreement of the parts with each other and with the whole, in the other (going back to Plotinus) there is no reference to parts but beauty is the eternal radiance of the 'One' shining through the material phenomenon. An analogous contrast can also be found later in the well-known quarrel between Goethe and Newton concerning the theory of colours: Goethe had a similar aversion to 'parts' and always emphasized the disturbing influence of instruments on the 'natural' phenomena. We should like to advocate the point of view that these controversial attitudes are really illustrations of the psychological contrast between the feeling type or intuitive type and the thinking type. Goethe and Fludd represent the feeling type and the intuitive approach, Newton and Kepler the thinking type: even Plotinus should probably not be called a systematic thinker, in contrast to Aristotle and Plato.

Just because the modern scholar prefers in principle not to ascribe to either one of these two opposite types a higher degree of consciousness than to the other, the old historical dispute between Kepler and Fludd may still be considered interesting as a matter of principle even in an age for which both Fludd's and Kepler's scientific ideas about world music have lost all significance. An added indication of this can be seen in particular in the fact that the 'quaternary' attitude of Fludd corresponds, in contrast to Kepler's 'trinitarian' attitude, from a psychological point of view, to a greater *completeness of experience*. Whereas Kepler conceives of the soul almost as a mathematically describable system of resonators, it has always been the symbolical image that has tried to express, in addition, the immeasurable side of experience which also includes the imponderables of the emotions and emotional evaluations. Even though at the cost of consciousness of the quantitative side of nature and its laws, Fludd's 'hieroglyphic' figures do try to preserve a *unity* of the inner experience of the 'observer' (as we should say today) and the external processes of nature, and thus a wholeness in its contemplation — a wholeness formerly contained in the idea of the analogy between microcosm and macrocosm but apparently already lacking in Kepler and the last in the world view of classical natural science.

Modern quantum physics again stresses the factor of the disturbance of phenomena through measurement, and modern psychology again utilizes symbolical images as raw material (especially those that have originated spontaneously in dreams and fantasies) in order to recognize process in the collective ('objective') psyche. Thus physics and psychology reflect again for modern man the old contrast between the quantitative and the qualitative. Since the time of Kepler and Fludd, however, the possibility of bridging these antithetical poles has become less remote. On the one hand, the idea of complementarity in modern physics has demonstrated to us, in a new kind of synthesis, that the contradiction in the applications of old contrasting conceptions (such as particle and wave) is only apparent; on the other hand, the employability of old alchemical ideas in the psychology of Jung points to a deeper unity of psychical and physical occurrences. To us, unlike Kepler and Fludd, the only acceptable point of view appears to be the one that recognizes *both* sides of reality — the quantitative and the qualitative, the physical and the psychical — as compatible with each other, and can embrace them simultaneously.

7

It is obviously out of the question for modern man to revert to the archaistic point of view that paid the price of its unity and completeness by a naïve ignorance of nature. His strong desire for a greater unification of his world view, however, impels him to recognize the significance of the pre-scientific stage of knowledge for the development of scientific ideas . . . by supplementing the investigation of this knowledge, directed inward. The former process is devoted to adjusting our knowledge to external objects; the latter should bring to light the archetypal images used in the creation of our scientific concepts. Only by combining both these directions of research may complete understanding be obtained.

Among scientists in particular, the universal desire for a greater unification of our world view is greatly intensified by that fact that, though we now have natural sciences, we no longer have a total scientific picture of the world. Since the discovery of the quantum of action, physics has gradually been forced to relinquish its proud claim to be able to understand, in principle, the *whole* world. This very circumstance, however, as a correction of earlier one-sidedness, could contain the germ of progress toward a unified conception of the entire cosmos of which the natural sciences are only a part.

I shall try to demonstrate this by reference to the still unsolved problem of the relationship between occurrences in the physical world and those in the soul ... modern science may have brought us closer to a more satisfying conception of this relationship by setting up, within the field of physics, the concept of *complementarity*. It would be most satisfactory of all if physics and psyche could be seen as complementary aspects of the same reality. We do not yet know, however, whether or not we are here confronted — as surmised by Bohr and other scientists — with a true complementary relation, involving mutual exclusion, in the sense that an exact observation of the physiological processes would result in such an interference with the psychical processes that the latter would become downright inaccessible to observation ... Furthermore, whereas older philosophical systems have located the psychical on the subjective side of the division, that is, on the side of the apprehending subject, and the material on the other side — the side of that which is objectively observed — the modern point of view is more liberal in this respect: microphysics shows that the means of observation can also consist of apparatuses that register automatically; modern psychology proves that there is on the side of that which is observed introspectively an unconscious psyche of considerable objective reality. Thereby the presumed objective order of nature is, on the one hand, relativized with respect to the no less indispensable means of observation outside the observed system; and, on the other, placed beyond the distinction of 'physical' and 'psychical'.

. . . in microphysics . . . the natural laws are of such a kind that every bit of knowledge gained from a measurement must be paid for by the loss of other, complementary items of knowledge. Every observation, therefore, interferes on an indeterminable scale both with the instruments of observation and with the system observed and interrupts the causal connection of the phenomena preceding it with those following it. This uncontrollable interaction between observer and system observed, taking place in every process of measurement, invalidates the deterministic conception of the phenomena assumed in classical physics: the series of events taking place according to pre-determined rules is interrupted . . . by the selective observation which . . . may be compared to a creation in the microcosm or even to a transmutation the results of which are, however, unpredictable and beyond human control.

In this way the role of the observer in modern physics is satisfactorily accounted for. The reaction of the knowledge gained on the gainer of that knowledge gives rise, however, to a situation transcending natural science, since it is necessary for the sake of the completeness of the experience connected therewith that it should have an obligatory force for the researcher. We have seen how not only alchemy but the heliocentric idea furnishes an instructive example of the problem as to how the process of knowing is connected with the religious experience of transmutation undergone by him who acquires knowledge. This connection can only be comprehended through symbols which both imaginatively express the emotional aspect of the experience and stand in vital relationship to the sum total of contemporary knowledge and the actual process of cognition. Just because in our times the possibility of such symbolism has become an alien idea, it may be considered especially interesting to examine another age to which the concepts of what is now called classical scientific mechanics were foreign but which permits us to prove the existence of a symbol that had, simultaneously, a religious and scientific function.

APPENDIX I

Fludd's Rejection of the Proposition that the Soul of Man is a Part of Nature

Replicatio in Apolog. ad Anal. XII
(Frankfurt on the Main, 1622), pp.20 f.

From these foundations of your *Harmonices* there arise, it seems to me, multifarious questions and doubts not easy to resolve, namely:

1. Whether the human soul is a part of nature?
2. Whether the circle with its divisions by the regular polygons is reflected in the soul because it [the soul] is an image of God?
3. Whether the determinants of the intellectual harmonies in the Mind Divine are established on the basis of the division of the circle which takes place in the essence of the soul itself, as Johannes Kepler would have it (p.21), whose model here is the human mind, which has retained from its archetype the impress of the geometrical data since the very beginning of man?
4. Whether the sense of hearing is a part of nature and bears witness to the sounds and their qualities as represented [to the intellect] by the *sensus communis*?
5. Assuming that (on the basis of the aforesaid) the proportion is reflected in the mind from its origin, whether then the sounds should be considered harmonious and whether pleasure can be derived from them or not?
6. Whether the triangle is a part of the nature of the intelligible things, likewise the square and whatever else divides the circle into parts which by their quantity or length determine any harmonious proportion, and whether all other natural factors that are present in

artificial song, follow the numerical value, so established, of the consonances?

On the main points of these questions, my Johannes, I shall begin to speak in order, not intending to contradict you in any way or to do any damage to your *Harmonices* but only for the sake of discussion and as a philosopher stimulated by another philosopher to solve some questions, quite apart from his own opinion:

Whether the human soul is a part of nature?

This question I must answer in the negative, contrary to what you hope.

1. Because nature, in its capacity of universal soul, contains the formula of the whole and is not even divisible into essential parts, as Plato testifies.

2. Hermes Trismegistus says that the soul, or the human mind (which he did not hesitate to call the nature of God), can as little be separated or divided from God as a sunbeam from the sun.

3. Plato as well as Aristotle seem to affirm that the Creator of all things possessed as soul something whole [total] before any division. And Plato called this soul universal nature.

4. Plato says that the soul, when separated from the corporeal laws, is not a number having a definite quantity and cannot be divided into parts or multiplied but is of *one* form [a continuum].

5. And Iamblicus seems to maintain that the soul, though it seems to have within itself all orders and categories, is nevertheless always determined according to some unity.

6. Finally, Pythagoras and all the other philosophers who were endowed with some touch of the divine recognized that God is one and indivisible. Wherefore we can argue syllogistically as follows:

   A. *That which was a whole before any division is not a part of something.*

B. *Now, the soul was a whole before any division.*
C. *Therefore it cannot be a part of nature.*

B is proved by the third axiom mentioned above. But if you say in objection to A that the philosopher meant the world-soul or the universal soul, whereas you mean the human soul, I reply with the fourth axiom that the soul separated from the corporeal laws is not a number and not divisible. Now that world-soul, which in Plato's opinion according to axiom 3 is nature itself, is separated from the corporeal laws. Consequently the human soul can also not be considered a part of the former since it is indivisible (as is proved by axioms 2, 3, and 4). Or I can deal with you in another way, by taking my argument from your own mouth:

A. *The image of God is not part of anything.*
B. *Now, on the basis of what has been granted, the human soul is the image of God.*
C. *Therefore, it is not a part of nature.*

A is clear because God is the One and Indivisible according to axiom 6. B is your own assertion as it is cited in the second question and as the speech of Hermes Trismegistus about the extent of the mind declares, according to axiom 2.

Now we shall go on to the second question: *Whether the circle with its divisions by the regular polygons is reflected in the soul because the latter is the image of God?*

I shall not hesitate to answer this question also in the negative, supported by the strongest and most encouraging arguments of the philosophers. Namely, because:

1. Plato, first of all, says that the soul, separated from the corporeal laws, is not a number having quantity and is neither divisible nor multiplicable. But it is uniform, revolving in itself, rational, and surpasses all corporeal and material things.
2. Aristotle and Plato say that the Creator maintained the soul as a totality before any division, and Pythagoras

makes it a 'one in itself' and says that it has its unity in the intellect.

3. Pythagoras, in his letter to Eusebius, acknowledges that God is a unity and indivisible and says that duality is the Devil and evil, because in it lies multiplicity and materiality. And Plato holds that all good exists as One; but evil, he holds, comes from chaotic multiplicity.

4. Cicero says that it would not be possible for perfect order to exist in all the parts of the world unless they were united by one single divine and continuous spirit.

5. God can neither be limited [defined] nor divided nor composed (according to Franciscus Georgius).

6. By the Platonic philosophers God is said to be present [*lit.* poured into] in all things. [He is called] the world-soul (which, they say, contains the formula of the whole) inasmuch as He, universally diffused, fills and invigorates all things.

7. God can be determined neither according to essence nor according to quality nor according to quantity, inasmuch as no predication can comprehend Him (Scotus).

8. The Pythagoreans and the Platonists regard the world-soul as being enclosed within the seven planetary spheres and say that within the first sphere it rests in the highest mind; and then, they say, it has become identical with it.

9. As all numbers are in the One, as all radii of the circle are in the centre, and as the powers of all the members are in the soul, so, it is said, is God in all things and all things in God. *Ars chymica.*

10. Hermes Trismegistus says that God is the centre of any one thing — a centre the periphery of which is nowhere.

With the help of these axioms of the philosophers I therefore argue thus against your assertions.

Argument I

A. *That which in and of itself is neither a number nor has quantity is not capable of receiving into itself any quantitative* [measurable] *figure (such as the circle).*

B. *Now the soul, which is freed from corporeal laws, is not a number and has no quantity.*
C. *Therefore the soul does not receive into itself from the very beginning a measurable figure (such as the circle); and consequently a circle is not in the least reflected in it.*

A is clear because a non-quantum [a non-quantitative magnitude] cannot receive into itself any quantities, as the One does not admit of multiplicity, and consequently is not a number.

B is confirmed by the first axiom and similarly by the second and third, according to which it is proved that the soul is one. But if you reply that the soul, as you conceive it, is not separated from physical laws since it is the human soul, then I say that you have meant the essence of the soul, as is apparent from your subsequent words; and in man, as he exists, this essence is not different from that soul of the macrocosm of which axiom 1 speaks in the second question, and according to axiom 2 of the first question where it is shown that the essence of the soul cannot be separated from God.

Or also thus:

A. *If the soul is an image of God it is neither a quantity nor a number.*
B. *Now it is, as you yourself admit, the image of God.*
C. *Therefore it is not a number nor does it admit of quantity.*

A is established because God, according to axiom 7, cannot be determined according to essence nor according to quality or quantity, inasmuch as he stands outside of and above any predication.

As far as the confirmation of your statement (demonstrating that the soul is the image of God) is concerned, however, this is also proved by axiom 7, which testifies to the fact that the soul rests at all times in God and becomes one with Him in the highest terminating sphere of its being. And [it is also proved] by axiom 2 in the first question according to which the mind is not divided from God.

A. *If the circle with its divisions by the regular polygons is (as you say) reflected in the soul from the very beginning, then the soul is divisible and multiplicable.*

B. *Now the soul is neither divisible nor multiplicable.*

C. *Therefore . . .*

A is evident because, if the circle filled it [the soul] completely (whence it is also designated as a circle by the Platonic philosophers, though only metaphorically speaking), and if this circle were divisible into parts by the regular polygons, it follows that the soul also would be divided by the divisions of that circle.

B is confirmed by axiom 1; furthermore it is shown clearly by axiom 2 that the Creator maintained the soul as a totality before any division, wherefore from the very beginning the circle was not reflected in it, nor did it admit of the divisions of the circle by the regular polygons. But this can be stated even more lucidly in the following argument:

A. *The human soul is (even in your assertion) an image of God.*

B. *Now God can neither be divided nor composed.*

C. *Therefore neither can the human soul.*

Replicatio, p.34:

. . . You maintain, then that the human soul is a part of nature and that in the soul the circle is reflected with its divisions by the regular polygons because of the fact that the soul is the image of God. But I say that the soul, at least with respect to its essence, cannot be divided from nature as a part can be divided from the whole, according to that statement of Hermes Trismegistus (*Poimandres* 12): The mind is in no wise divided from the essence of God. Rather it is bound up with Him as is the light with the body of the sun. For we see that the solar rays are bound up with the body of the sun and cannot by any means really be divided from it, because the essence of light is a unity and cannot be divided into parts; but naturally with respect to us who abide in multiplicity we say that the soul of one man differs from that of another in number and kind, although in truth all souls have a continuous relation to the *one* world-soul or the Metattron, as has the sunlight to the sun. Consequently, multiplicity really lies in matter and not in form which is nothing but a continuous emanation from God, or the Word of God, imparting life and being to all creatures. When it is withdrawn [revoked] life is destroyed, as it says in Psalm 104 . . .

Replicatio, p.35:

I therefore conclude that, as God's essence is indivisible, so also nature itself, which is His emanation into the world, is in every respect *one single form* and indivisible in itself. And [only] in so far as God — and, hence, the functions and qualities produced in order to perfect the world — is divisible into three Persons, [only] in so far does one say that the soul, too, can be divided into various parts, whence it is sometimes the senses, now memory, now imagination, then reason, intellect, mind, and so on.

Those, then, who seek to consider the soul as it inheres in perishable things will observe with their physical eyes that is can be distinguished from the body and it properties. But he who, turning back into himself and to his centre and neglecting the external world as a deceitful shadow, penetrates into his inner gateways, he will perceive with his spiritual eyes that there is neither divisibility nor quantity in the soul and that neither numbers nor geometrical figures can be discovered in God (Who is above quantity and quality, Who has a continuous essence of soul) . . .

But the world-soul is not on this account a circle, neither is there a circle within it; but rather by its own circular motion it encompasses and contains the universe as in the most capacious figure, and also divides it from the darkness of matter. The circle and its imaginary divisions exist, therefore, in the created passive spirit and not in the creating soul.

APPENDIX II

Fludd on the Quaternary

Demonstratio quadam analytica (Discursus analyticus)
(Frankfurt on the Main, 1621), Analysis of Text XXI,
p.31

Here the dignity of the quaternary number will be discussed and I shall defend it with might and main as far as my weak intellect allows, spurred on by the insolence of the author [Kepler]. Not only has sacred theology extolled the paramount superiority of this number above the others, for which reason I feel myself moved to regard and acknowledge it as divine; but also Nature herself, the maid of the Godhead, and the nobler mathematical sciences, that is to say, Arithmetic, Geometry, Music, and Astronomy, have demonstrated its wonderful effects. Hence, when we examine thoroughly its praise in theology, we shall perceive, first of all, that this quadratic number is likened to God the Father in whom the mystery of the whole sacred Trinity is embraced. For the first and simple proportion of the quaternary, that of 1:1, denotes the symbol of the monad, the super-substantial essence of the Father, proceeding from which the second monad engendered the Son like unto Itself, and this second progression is also simple, as 1 from 1. The proportion of 2:2, which is the second progression from the simple numbers, denotes the Holy Ghost, proceeding from two, namely from Father and Son. These progressions in the quaternary are lucidly expressed by the ineffable name   [Jahweh]: where the double He or [h] signifies the progression from Jod the Father and from Vau the Son, wherefore this name [that] alone expresses the essence of God and no other is known as *Tetragrammaton*. And this is the reason why this number is called by the wise the *Origin and Source of the whole godhead*. Nature herself, deriving her origin from the Godhead, also lays claim to this number as to her fundamental principle. And this is the same thing which the Pythagoreans proclaimed who called this number *the eternal fountainhead of nature*, as appears from the following verses which the Pythagoreans were accustomed to pronounce when taking an oath:

> *Pure in heart I swear to thee by the holy Four,*
> *the fountainhead of eternal Nature, the procreator of the soul.*

And while I am discussing this subject I shall say the following: the Pythagoreans did not consider duality as a number but as a blending of the unities. Consequently they declared its square to be that first even number, and not without reason; for, since the first unity signifies the divine form or *actus*, the second unity, however, the divine *potentia* or matter: *potentia* must needs emerge from darkness by virtue of the *actus*. Of these unities, now, the first was created, through the binding action of the three-fold Unity, from the general [unspecified] substance of the world according to the nature of the holy Trinity. But because the first square was based upon the number 2, the progression of nature proceeded to that number 4 whose proportion to the number 2 is 2:1, and which is also the square of the number 2. And in this way the general, watery substance of the world was divided into four elements distinct from each other. From this number [4] there is in the order of things a progression equal to the first, namely, to the first cube which is the number 8 or 2 x 4; this denotes the compositions of the elements just as the elements themselves, like the square, proceeded from the number 2 which denotes simple matter and simple form as distinct from each other. From this, then, there originated the four degrees of nature which are related to the four elements, namely, being, life, sensory perception, and intelligence; the four cardinal points of the universe; the four triads in the firmament [that is to say, the four groups of three zodiacal signs, each corresponding to one of the seasons]; the four primary qualities beneath the firmament; and the four seasons. Indeed all nature can be comprehended in terms of four concepts: substance, quality, quantity, and motion. *In fine*, a quadruple order constantly pervades the entire nature, namely, seminal force, natural growth, maturing form, and the compost. By this we can clearly demonstrate that this number 4 should rather be chosen to distinguish and divide the humid [primal] matter than the number 3 or the number 5.

Arithmetic also demonstrates the superiority of this number to all others. For, this science well explains not only its twofold proportion (first, that of 1:2, second, that of 2:4) but also the origin of this proportion in that it is produced by and born from a twofold progression and proportion, namely, from 1 to 1 and from 2 to 2. The number 4 thus begins with the unity and ends in the quaternity. And indeed in this number all others are contained, for $1 + 1 = 2$, and $1 + 2 = 3$, and $3 + 1 = 4$. Thus then, are established 1, 2, 3, and 4, in which all the mysteries of the whole world and Nature herself and the extent of arithmetic are contained; for, by 3 and 4 is produced the number 7 which, *formaliter* [symbolically] considered, is downright mystical and full of secrets. From the addition of 2 and 3 there results the number 5; from $1 + 2 + 3$, the number 6; from $1 + 3 + 4$, the number 8; from $2 + 3 + 4$, the number 9; and, finally, from the summation of the entire natural progression $1 + 2 + 3 + 4$ there arises the number 10, beyond the designation of which there can be no more progress. From these progression there arise all proportions of geometry and music, as 1, 2, 4, 6, 8, 10, and 1, 3, 6, 9, and 1, 4, 8. And from him who properly understands the use of this natural progression 1, 2, 3, 4 in formal [symbolical] speculation there will not be hidden the mystery of the seven days of creation; and why the sun was created on the fourth day; and how $3 + 4$ constitute either the 7 or the 10 or the 4 among the rational numbers; and why the number 4 is the number of the day of Sabbath, viz., rest; and why the number 4 is the day of the sun; likewise, how in the true operation of nature the triad denotes and establishes the hexad and brings about six days in the work of creation, he will also be able to work out the formulae of the Critical Days and the Climactic Years. When he considers the 4 as a unity he will see, with open eyes as it were, the creation of the seven planets in the world, and many other wonders.

In geometry its power is infinite inasmuch as it comprises this part of mathematics in four concepts: point, line, surface, and body. From it [the number 4] we also see emerging that aboriginal geometrical cube from the innermost part of which our author Kepler has produced all the just as the four elements from the womb of chaos; for, the cube, which he himself acknowledges as primordial and containing the formula of everything, results from the multiplication of the square. Since this is so, I was obliged to choose in my divisions the number 4, into which the cube can be resolved as into its primary elements — namely, squares — from which, by his own admission the triangle and the pentagon are obtained. Consequently, a composite natural thing, related as it is to the cube, should be divided into quarter, viz., squares, rather than into three thirds or five fifths. For, in the act of decomposition there takes place the dissolution of the composite, viz., the cube, into four elements, that is to say, into the square; just as, conversely, in the act of generation there is a natural progression from the square to the cube.

Finally the power of this number is revealed as clearly as possible in the science of music, inasmuch as it comprehends in itself the entire musical harmony. For in the double [proportion], as 1:2, lies the octave; in the sesquialtera, i.e. 2:3, the fifth; and in the sesquitertia, i.e., 3:4, the fourth. Furthermore, from the number 4 and its root there result all the proportions of the composite consonances [chords]. The octave, for example, stands in relation to the fifth in the triple [proportion], i.e. as 2, 4, 6. For, between 2 and 6 a triple proportion is assembled from the double, namely 2 + 4, and the sesquialtera, i.e. 4:6. The double octave is found in the fourfold [proportion], as 2, 4, 8; the fourth, however, plus the fifth makes one octave, as 2, 3, 4. From this it can be seen that all musical proportions receive their properties from the quaternary and its root and either resolve themselves into its measures or arise from them.

And, finally, if we consider mystic Astronomy we shall indeed perceive in it the whole power of the quaternary, and this most clearly; for its whole secret lies in the hieroglyphic monad which exhibits the symbols of the sun, moon, the elements, and fire, that is to say, those four which are actively and passively at work in the universe in order to produce therein the perpetual changes whereby corruption and generation take place in it. [See Figures 4-8, pp.99, 171-6.]

In its symbolic image we see, first of all, an indication of the quaternary in the cross, four lines being arranged so as to meet in a common point. Joined with the number 3, which denotes the moon, the sun, and fire, this [quaternary] will produce the number 7, which can also be demonstrated by the four elements. And yet this number 7, is in itself none other than the quaternary considered formally.

Furthermore, even the practitioners of ordinary astronomy have esteemed this matter as of great moment: in establishing the Zodiac, they divided it into four triads. We conclude, therefore, that the wise men called this number *Tetraktys* and gave it precedence above all other numbers because, as has been said, it is the foundation and root of all other numbers. Hence all fundamentals, both in artificial and natural things, and even in the realm divine as well, are squares, as has been explained above. It follows, therefore, that the division of a natural thing by the number 4, which in the order of nature itself, is preferable to a division by the numbers 3 or 5, which are by nature derived from the root of the quaternary and consequently subordinated to it. Finally, in dividing the earth into four parts, the water into three, the air into two, and the fire into one, one should not understand this distribution as the author [Kepler] does, as has been expounded above but with respect to the formal proportion in those elements. For I endeavour to show that the nature of the earth, since it is the basis and, as it were, the source and cube of matter, has little or nothing of form or vivifying light in itself; it is, so to speak, the vessel or matrix of nature and the receptacle of the celestial influences, so that the light that it has belongs to it more by accident than by nature, inasmuch as it [the earth] is very far removed from the source of light and is the coldest of all elements, and this in the fourth degree; water is also cold but to a lesser degree. For this reason it [the earth] admits of only one degree of light into itself and so also in the case of the others. The wise ought therefore to understand rightly before condemning rashly.

Notes

1. This paper was published with a companion essay by Carl Jung, 'Synchronicity: An Acausal Connecting Principle' in *The Interpretation of Nature and the Psyche*, Bollingen Series 51 (Pantheon Books, New York, 1955).

   An essential follow-up to Pauli's essay is Robert S. Westman's 'Nature, Art and Psyche: Jung, Pauli, and the Kepler-Fludd polemic' in *Occult and Scientific Mentalities in the Renaissance*, ed. Brian Vickers (Cambridge University Press, Cambridge, 1984).

   Westman argues that Fludd's *picturae* go beyond being illustrations for his texts: they are 'ways of *knowing, demonstrating, and remembering*'. They help the viewer direct the self back to inner unity with the Creator. The human form provides the iconography that bridges the gap between the theoretical and practical, because it includes both geometric principles and symmetry. Westman believes that Albrecht Dürer was Fludd's source for this scheme. Fludd refers to Dürer's *Four Books on Human Proportions*, and borrowed several illustrations from it, most particularly the key one of the harmonic proportions of man, the microcosm.

   Westman shows convincingly that the 'true theme' of the Jung and Pauli essays is an attempt to formulate a resolution of the seeming opposition between the quantitative thinking view and the qualitative intuitive one, of which Fludd and Kepler are examples. He also reveals the fascinating discovery that it was Jung's interpretation of Pauli's dreams that Jung used to formulate his concept of the mandala as the symbol of the self. The dreams, by a previously unknown subject, were published in Jung's *Psychology and Alchemy*, where they are accompanied by many illustrations from Renaissance alchemical texts. The illustrations were meant to show that alchemical symbolism reflects psychic states described in the dreams. The lead illustration at the beginning of the chapter called 'The Initial

Dreams' (now known to be Pauli's) is one of Fludd's — 'The Mirror of the Whole of Nature and the Image of Art'. Jung gave it this caption: 'The *anima mundi*, guide of mankind, herself guided by God.' Later in the book, Jung maintains that 'the *anima mundi* [Fludd's illustration is cited] coincides with that of the collective unconscious, whose centre is the self.'

Finally, Westman concludes: 'Pauli must have believed that Fludd's pictures represented symbols of the collective unconscious and the self, and that by studying Fludd he was gaining access to the *Fluddean part of himself*' since he was experiencing this same conflict of visualization in quantum mechanics. Modern physicists representing this conflict may be seen in Heisenberg's pure mathematical formulations as opposed to the imaging of Schrödinger or Bohr.

2. The weather-glass actually first appeared in Fludd's *Integrum Morborum Mysterium* of 1631; the *Philosophia Moysaica* was published in 1638. Kepler died in 1630.

The Efficacy of Alchemy: 'Truth's Golden Harrow'[1]

Truth's golden Harrow
Framed and fashioned by a loyal (and
faithful)[2] servant of Virgin
Alchemy
To break and prepare those gross
clods of Error (and to correct
those furrow balks) wch the
blunt Plowshare of a mis-
Believer in the material
Elixir hath left im-
perfect in his Till-
age of Light.
*Latent adhuc Lilium et Rosa
inter spinas.*
Experientia Veritatis sigilum.

Gentle Reader,
    If our late Tiller of light has rudely passed over and but
slightly performed the weighty charge and important task
which with so great a show and demonstration of courage and

sufficiency he did undergo, in mine opinion, he is not altogether to be condemned for [being] insufficient, though on the other side not to be excused for his oversight and negligence in leaving behind him many irregular balks of error unrectified and diverse clods of confusion unbruised and ill prepared. But his confession Page 32 hath in some sort mitigated that offence, where he acknowledges that his plow ... [has but] shortly gone over the large field of the philosophers Elixir. Now haste you know breeds waste: *Festina lente* sayeth the philosopher, and *Lente lente et omnia bene*. To pass over so great a field as that of the philosophical Elixir with so slight and short a plow toil must either produce a very shallow tillage or cause a rugged and ill manured soil.

Lo therefore I have imagined it a thing most convenient, to provide a harrow framed out of philosophical gold, to correct those balks and unleveled paths which the plow hath made in this tillage, whereby the ground may become more apt to receive the grain of reality and truth, which before was only bestrewed with the appearing seed[s] of morality which are wondrously annoyed with the pricking thistles of contradiction and mingled with the unprofitable darnel of obscurity. This I say, is all, and this all I freely recommend unto your wise and impartial censure to be fully determined.

Farewell

[Fludd summarizes Scots' points and breaks them down to 12 'furrows', which he 'harrows' in turn.]

The First Furrow

[Scot maintains that he wishes to forestall others from the unprofitable pursuit of a nonentity, the Philosopher's Stone or elixir, or artificial gold or light incorporate by art. Fludd's reply is as follows:]

Alas, how many wooers has this golden Elixir had, who by their best endeavours have laboured and toiled to purchase but a view of her bright countenance, and to enjoy but the weight of a grain of mustard seed of her grace and perfection; all which, perceiving themselves at the last either neglected or else altogether rejected of her as being far unworthy of her favours, have turned their coats, and in lieu of her wanted commendations have contaminated her virgin purity and supernatural simplicity with abominable lies, marked her irreligiously in the forehead with the brands of disgrace and ignominy, proclaiming her a sophisticated strumpet, a chimerian and imaginary *non ens*, a smoke without a substance, a void character of what is named but has no essence, affirming therefore that she is not because she will not deign to reveal herself but unto very few, and those must prove worthy of her graces and favours. For such therefore and unto such was she sent down from heaven, with them she delights and dwells, a few she embraces and hourly visits with comfort; a million she rejects as strangers, and therefore she casts before their sight a mist of darkness and error that she might not appear unto them, but that they might have eyes and not see and so might live in perpetual darkness.

To conclude I suspect that this our Tiller of light will prove but a sophisticator and one who after long suit has been rejected and refused by this peerless Lady, the philosophers Virgin Elixir, who, because he has been and yet continues so blind that he could never purchase with all his endeavours a sight of her bright presence, does imagine that she is not [to] be found *in rerum natura*, for it seems that he presumes so far on his own worth and merits that it should stand for a position already granted, that of necessity she must reveal herself unto him if there were any such thing, where as contrarywise the humble minded philosophers do and will confess that more mysteries are hidden from the sense of man than the things are which are revealed unto it.

But me thinks that our harrow, being ready to conclude the preparation of this furrow, lighteth . . . on a certain balk of contradiction. For in our husbandmans charitable affection towards poor distressed Alchemists, in being a feeling copartner in the miserable delapidation of their means and fortunes [from the search for] an imaginary Elixir, he seems altogether to vary from the rules and precepts of those philosophers doctrine which he so earnestly teach[es]: he says and alleges that the philosophers by their so punctually setting down a seeming real Elixir had a good intent, being (as he surmises) to exercise curious spirits lest they should precipitate themselves upon more dangerous rocks of higher forbidden mysteries, or become alltogether idle . . .

Now the philosophers knew that it was impossible to practise the making of the Elixir after their enigmatical and parabolical doctrine without much expense and charge . . . the ancient philosophers do teach us that . . . poor men ought not to be employed in the business of Alchemy; so the wisest and plainest among them do ingeniously conclude that *quicquid facis tolle manum a marsupio*: What so ever thou doest (say they) put not thy hand into thy purse. For our art consists not in the multitude of expenses.

But suppose that by the practice of this science men were brought unto poverty and necessity: our author confesses that such were the lives and conditions of the philosophers whose custom it was to condemn gold to the intent that by poverty they might with the greater facility attain unto the wise mans *summum bonum*, which was the main butt at which they aimed. I wonder therefore why such extreme love of poor distressed Alchemists should move our Tiller of light to withdraw them from that which the wiser philosopher (as he thinks) invented of set purpose to keep and withdraw them from greater vices. Or why should he counsel men to preserve so warily their wealth and estates which he so exclaims against, saying, O deceptful riches, how falsely are you called goods, riches and gold, I say, which the wise philosophers so condemned and rejected, as he himself does infer . . .

The Second Furrow

[Scot exclaims that] This artificial gold, imaginary Elixir . . . is a poisonable pill gilded with sophisticated curiosity, base covetousness or encroaching, cunning, emulous strangers and irreconcilable enemies unto philosophy.

[Fludd replies] If the Philosophers Stone [is] such a masked, sophisticated and prestigious a thing as he makes it, then will it without all doubt by effect show itself in his colours as an open enemy unto philosophy.

I confess that the philosophers do jointly agree that it is as venomous as a viper or adder in the working, and therefore we are warned to beware of the breath of it, but likewise they all so assure us that in the conclusion . . . it becomes a perfect medicine of life, and the perfect *aurum potabile* which cures the leprosy and preserves the body from diseases, yea and rectifies the spirit of man . . .

Also I must acknowledge with the philosophers that this pill is gilded over with sophisticated curiosity in the eyes of fools and such as are ignorant, but to the wise and understanding person it proves a true Elixir of Life . . .

In like manner unto such as go about to seek it . . . that is unprepared and with the filthy pretence of lucre and gold only, unto such persons it leaves the title of baseness, and will not easily be found by such, but those which zealously seek after it and with a true intent, and by Gods blessing do attain it, shall easily perceive that gold is the meanest possession and the least to be esteemed of ten thousand other mysteries which it brings with it, and this was the reason that the philosophers did esteem gold but little being that it was the unworthyest blossom of so divine and precious a plant.

Lastly where he says that it is an emulous stranger and irreconcilable enemy unto philosophy, I would trust him a little if it were but a chimera or *non ens* (as he tells us) but it blasts his cunning to prove it, being that we have more authentic authors to confirm the reality of it, for we find that both philosophy and divinity do accord in the essential existence of the *summum bonum*, this multiplying light, this cupido of nature, this Cabalists Metattron or Platonist universal soul of the world, by which the effect of this creating word, *Crescite & multiplicamini*, is produced into act in every creature of what kind so ever.

The Third Furrow

[Scot states that the elixir is an enemy of philosophy.]

[Fludd: T]his Elixir is the true temple of wisdom, the impregnable castle of Cupid, that powerful god of Love, the beautious and bright city of the sages, the true pattern of the heavenly Jerusalem, the mark of perfection at which all imperfect spirits do tend as to the port of their final happiness, the scale of Justice, the queller and extinguisher of vice, and the final complement and exaltation of form, and exact being. And therefore he that is not a true believer and lover of this excellent masterpiece, is drowned in darkness and has an iron gate before his eyes of understanding, for this is the mirror of truth, the clear Thummim of the ancient Jews in whose centre dwells their bright Urim as a divine soul in an unpolluted and chaste and virginal body, the gold of God, the gold that is to be bought of Christ, the guider of mens actions, the house of wisdom propped up with 7 pillars.

And to conclude, instead of being an enemy to philosophy, it is the greatest friend she has; for the only Lady she serves, as the body does the soul, is wisdom, which is the *summum bonum* of the philosophers, and main subject of philosophy who therefore has her denomination from her love [of] wisdom. Now the Elixir is the temple of wisdom, or the earthly sun of the philosophers which is as well the tabernacle of the divine emanation as the heavenly. The earth shall open and bring forth a saviour. Light is in darkness and darkness does not comprehend it.

The Fourth Furrow

[Scot asserts that in holy writ wisdom was given in mysteries, parables, allegories and analogies, but that it is not warranted to say they applied to a material elixir.]

[Fludd:] In Holy Scriptures we find this allegory mentioned: 'The silver surely has his vein, and the gold his place where they take it, iron is taken out of the dust, and brass is molten out of the stone, God puts an end to darkness and he trys the perfection of all things; He sets a bound to darkness and the shadow of death. Out of the same earth comes bread [in margin: Job 28.] and under it, as it were, fire is turned up; the stones thereof are a place of sapphires and the dust of it is gold.'

A more excellent description of the material Elixir cannot be made by the wisest Alchemist or deepest philosopher, for the Agent in this work shining out of darkness (which is the Earth) is the fire or central sun, the patient is the elementary substance appearing out the said darkness and by purification and rotation of [the] element reduced to a spiritual rock or pure transparent sapphire, the effect is quick gold, the form or divine soul is the light shining out of darkness, the matter is the Earth, the intermediate spirit the rock of an azure or celestial colour which argues the quintessential spirit, again, the body is the earth refined into the powder of lively gold, to which perfection all the earth shall be reduced at the latter day, as by scriptures we are warranted, where we find it spoken of a new heaven and a new earth, and again *Ecce omnia nove sunt facta*: Lo, I make all things new. Ezekiel speaks of a fire that issued out of the dark cloud which came from the north, and out of the midst of the fire came, as it were, the likeness of amber, and the similitude of the firmament was as crystal: *Venit aurum ab aquilone*: of such an apparition also speaketh the Revelation [in margin: Revelat. c.4].

So we may discern in the place before mentioned in Job as well the natural [in] the generation of metals as that which is made supernatural by the administration of art. For art is ordained in this mystery to the intent that by the adaptation of kinds rightly, light might shine out of darkness by the violating of the bands thereof, which having liberty of action works on the deformed substance and leaves not until it has reduced it to the highest period or exaltation of act and plusquamperfection, for it leaves not his operation until it has of duality made unity, so that as out of one fountain of light two issued and were compounded, namely matter and form, so by progression into trinity, duality (the author of discord) might again be reduced unto unity. This is therefore the true type whereby we may be persuaded that by the saviour which issued from the dark earth, by conquering of death and darkness, we shall also (as Job teaches us) in this very body, being by light first purified, have a complete resurrection of body and soul unto eternity . . .

We cannot deny but that Christ the author of salvation (whose image and pattern this our mystery is) did rise both body and soul and so of two united together in perfection made one unity, transmuted darkness into light, mortality into immortality, and so made his passage from Unum which is the beginning to *bonum* or felicity which is the end, and these are both convertible and one only thing by the connection and unity of one spirit which is all one with them both. What, shall we therefore imagine him not to be because [he is] not material, or shall we deem him not material because spiritual, when every form imparts a matter, be it corporal or spiritual? Admit therefore that the Elixir be of a spiritual substance, [and that] excludes [not] it[s] . . . materiality, when corporality and spirituality vary not but in the refining and purifying, for . . . as Hermes says and St Paul witnesses [in margin: Ad Hebr. 11], All things visible were made of those things which were invisible by the word of God.

And St Peter [in margin: 2 Pet. 3] agrees with them both in this point saying, The heavens and the earth were of the water and by the water by the word of God; and Moses says the spirit of the Lord moved upon the waters. Now it is certain that this word and spirit of God is that first begotten wisdom the which dwells in the universal waters of this world as in the humid tabernacle of the created Nature and consequently, where so ever the bright beam of wisdom appears manifestly, there must the watery tabernacle be pure[ly] spiritual and virginal, the which nevertheless as it is fashioned of water of what degree of refination so ever must be material, and in respect of his perfection in maturity is rightly termed Elixir, as Thummim amongst the Hebrews signifies perfection in whose centre does that supernatural and plusquamperfect light Urim dwell as the form of Bright amber did in the bowels or midst of the fire which appeared unto Ezekiel out of the cloud [in margin: Ezek: c.1].

We must therefore on these grounds conclude that if all the mysteries, parables and oracles of holy writ alluded to such a wisdom as the spiritual rock . . . which is Christ risen again, [and is] composed of a divine spirit and a spiritual body, of which the true philosophers Elixir is said to be the type or pattern, we must not nor cannot justly affirm that this divine and spiritual stone can be excluded from materiality . . . It consists of a divine and plusquamperfect spirit and a body exalted from corporiety into a pure and spiritual existence, from mortality into immortality, and being the pattern of Christ risen again, it must needs have the power to multiply infinitely, according to that saying of Christ: When I am exalted I will draw all bodies unto me.

Neither let it seem strange to any man that such power is given to this light, being that it is a branch of the universal emanation from the fountain of light which was . . . present with God when he made all things; and thereupon Orpheus, Democritus and many of the Pythagoreans did imagine that all things were full of gods, to each particular of which they observe several venerations, prayers and sacrifices, and yet nevertheless such was their respect to the fountain of all these petty lights that they had ever more their main relation to one monache or sole Jupiter. By the same reason also the Platonists did acknowledge a peculiar soul or beam of light to be in every particular and individual creature, but the universal soul of the world was the head and fountain to which they referred all those singular lights as all the beams of heavenly light are issuing and belonging to one Apollo, one sun. Do not their opinions jump and meet in one point with this assertion of Holy Scriptures: *Spiritus Dei incorruptibilis est in omni re*: Gods incorruptible spirit is in everything.

This wisdom or first created nature . . . seems to be that invisible fire of Zoraster and Heraclitus of which all things were begotten; and it is said to be the spirit of the Lord which was carried on the face of the waters, which to them was in stead of the soul to a body, and this fiery spirit St Au[gu]stine calls *amorem igneum*, the fiery love which did impart a vivifying and vegetating vigour and force to the general waters or spirit of the world, being that without natural heat no generation or procreation could be had in *rerum natura*; for God, when he said in the creation, Increase and multiply, did inspire into every creature a certain germinating and vegetating spirit or viridity by which all things did multiply and increase their kinds. And this multiplying spirit was bestowed as well on minerals as either animals and vegetables . . .

. . . I will refer you to the sense of the 17th Hymn of the reverend Bishop Synesius . . . Now the divine Mens or bright soul and mental beam [only relates to] the intellectual world, and from this disposition the soul and reasonable spirit of man is derived. But this mental beam, being the offspring from immortal and divine parents gliding down into the dark Hyle or chaos, very small in substance, and yet nevertheless being all and one everywhere dispersed in the world; turns about by her power and virtue the vast and wide cavity of the heavens, and preserves them from ruin and corruption by her presence, for she is everywhere present by changing and fashioning herself into diverse forms, for part of her is employed to give motion and life to the stars, part institutes the order of the angels, and again part [imbues the] elementary and earthly shape which reciprocally embraces, with a grievous tie or knot, in so much that she, being separated from her immortal parents . . . sucks in dark oblivion, and so forgetting herself she admires the unpleasing earth, respects it with a blind solicitude and care, and by that means is prone to affect corporal things, and to incline itself to human affairs.

But it is manifest that those spirits which are thus descended and included, have in them a virtue which is able to call them back again to heaven, so that after, by virtue of it, they have escaped the troublesome waves of this life and go to that holy passage which directs to their kingly parents. Happy is he who, eschewing the devouring jaws of hyla [that] issues out of darkness, and with a light skip, directs his course to God. Happy is he who, after death, after labours and bitter cares of the earth, by entering into the ways of the mental beam, sees the altitude of things shining and glittering with divine light. This far passes this bishops discourse wherein he seems to admonish us by the type of this supernatural progression in creation, of the descent and final resurrection not only of man, but of all other creatures whatsoever in which this perfection of original light is sowed.

Did not St Paul make the grain of wheat his pattern of resurrection to the heaven of perfection or highest effect of multiplication? Now this beam or bright spirit of light inhabits this grain or else it could not rise again, neither could that spark have been set at liberty but by the adaptation of the husbandman, and the effect of an external nature working of the composition to putrefaction, that by that means the bands of the elements being so violated, the included spark of light might be set at liberty and consequently might have the better scope to work after his appetite, which tends to a perfection which is like itself. By this therefore it appears that there is a material substance that belongs to this infuse spark of wisdom.

Again by holy writ we are warranted that the essence of God ... fills everything in heaven and in earth ... [and is] attired in a natural or material vestment ... for in the highest heaven he is indued ... with light as with a vestment, and in this light he dwells centrally; then in the middle heaven he is said to have planted his tabernacle in the sun, from whence he liberally disperses everywhere his multiplying graces; moreover, in the elementary heaven he made the dark clouds his dwelling-place. In the lively earth of man he erected his spiritual temple [in margin: Sap. (Wisd.) 1. Psal. 139.7; Psal. 104.2; Sap. 12.1].

To conclude, his incorruptible spirit is in everything. All which being manifestly confirmed by Holy Scriptures, why should we doubt, much less deny, that the spirit of wisdom (which is the spirit and word of God) is housed in visible and material creatures, and consequently why should it not be the soul and multiplying life of the exalted matter of the Elixir, being that where this bright spirit finds a matter aptly disposed, and dignified in the highest degree according to the nobility of her divine form and essence, it will have dominion over darkness and shine forth as it doth out of the pure body of the heavenly sun, and bestow her graces out of the little world here on earth among men, as it does out of the sun of heaven in the great world. This is the meaning of Plato by this maxim of his: According to the merit of matter a more or less perfect form is given or bestowed on it from above.

God is called of holy writ a consuming fire, and we see that his fire is multiplicative *in infinitum*, by reason of that fiery spirits act within it, and yet we find that the flame differs according to the purity of the matter. For the fire of coal differs from that of wood, and that again from the candle, and that of the candle from Terpentine, and that again from brimstone, and all these from that which in the spirit of wine which by reason of his purity . . . We infer therefore . . . that it is . . . of little effect . . . that because the philosophers hieroglyphics and the Theosophists mysteries and parables did principally point at wisdom . . . they should not respect any material Elixir. As who should say, because the Holy Bible did aim at God, therefore they respected not the fiery bush out of which He appeared to Moses . . .

In a word, because Jehova was the guider of the Israelites in the wilderness, therefore he did not make use of material instruments, as a pillar of cloud and fire, for in that pillar it is written that Jehova, or the angel in which his great name was written, did serve as a bright soul unto a spiritual, but material body. Whereby it is evident that though wisdom be the mark the Philosophers and Theosophist[s] do aim at, yet they consider her as she is in a created nature as well subject to the sense as [the] invisible, and consequently material, for to speak of God and His spirit as it is without all things and within nothing, it were a great folly in man being that mans wit could never discern him *a priori* but by the effects of his dearest creatures. Though Moses saw his posteriors, yet he never beheld his essential face or being; wherefore the Cabalists confess him to have attained to the 49th gate of intelligence, but to the 50th his human composition could never attain. Thus we have our proofs of a material Elixir of perfection out of holy writ.

Let us now see how the sage Cabalists do agree with this doctrine of the sacred Bible. Let us hear what their opinion is of this first created light, which Moses termed the work Fiat or the Spirit of the Lord which walked on the waters by which all things were made. And St Peter affirmeth [in margin: 2 Pet:3] that heaven and earth were made of waters and by waters by the word of God, where he makes waters the matter or patient and the bright word the form or agent which accomplished and brought to perfection the world consisting of heaven and earth. And Moses said in the like sense, the Spirit of the Lord moved on the waters. And St Au[gu]stine says that the fiery Love gave a vivifying and multiplying vigour unto them. And Aristotle acknowledged a primary form which he called the first Act which did universally inform all matter. And Plato calls it the soul of the world which he measures by 999; for three times 9, amounting to 27, makes the cube of the root 3, which is the most perfect number and therefore attributed to the soul or first act in every creature. As 2, which is the number of confusion (as Pythagoras says), is the root of matter whose square in 4, and therefore his [cube] is 8. This 999 of Plato, by the addition of the Cabalist Aleph which signifies 1 in Arithmetic, makes up 1,000, beyond the which there is no denomination; and therefore as Aleph was one, and consequently the beginning, so also is it that one which is the end of all things . . .

We may observe . . . that there is required [for] the Elixirs perfection, matter and form. The matter is the passive or material substance which serves [as] a vehicle to carry and possess the form. But as the purer the form is, so also it requires a more pure substance or matter to work upon, so is it most aptly to be conceived that the most perfect and highest formal light requires a spirit to dwell in like himself. And therefore as the external beam of gold is made comfortable to the fixe[d] and purified matter in which it dwells, and therefore aspires to the common perfection alloted to gold in his creation, so if the subtle material spirit which dwells in that grosser fixe[d] matter could be had, in it shall we find that excellent formal light in his greatest activity which will, being thus at liberty and in a spirit obedient to his will, work of itself without any other manual application, the spirit which includes it to the plusquamperfection of itself and so causes a resurrection after the passive spirit is made all one with his agent, so that this matter will be transmuted into the nature of incorruptibility and immortality, the which is able to raise up other bodies by exaltation, according to that saying, When I am exalted, I draw all bodies unto me.

But will you yet in a more succinct file of speech understand what this spiritual vehicle, of the mental beam, or this material temple of the incorruptible spirit of wisdom is, which both Theology and Philosophy do so often mention in their discourses. Listen a while and I will first describe to you this house of wisdom, and then afterward I will show you briefly what wisdom itself is . . .

The external Elixir therefore we define to be a spiritual body made worthy by the action of Nature and the assistance of Art to receive so excellent and supernatural a shape or formal light as the spark or beam of wisdom is. And this exalted body is the true pattern of the perfect and spiritualized body of Adam in his innocency, in which no darkness was to be found but absolute simplicity in matter, even such a substance as we expect to have when we shall be entered into the gate of felicity. It is therefore the crystal palace of fiery Cupid or divine wisdom which is built on 7 pillars as Solomon teaches us, the enchanted castle of light and perfection into which we must find an entrance if we would walk in the pathways of light . . . It is the fiery bush of Moses and Esdras out of which God did speak to them, not permitting them to approach to it until they had put off their shoes, saying that they trod on hallowed ground, meaning that they should purify themselves. It was the fire of Ezekiel which issues out of the dark cloud, in whose middle appeared the likeness of amber. It is the celestial Jerusalem whose walls are made of precious stones, and whose gates are 12 in number, in the which glorious city the Bright Lamb is said to dwell and give a greater light than either sun or moon or stars, after the pattern whereof the Epod of Aaron beset with 12 precious gems resembling the 12 gates of the celestial Jerusalem in the midst of which Urim and Thummim or light and perfection were placed resembling the Lamb and His brightness in the midst of the said city. And unto this mystery has [George] Ripley chemically alluded [in] his [*Book of the*] *12 Gates*, and *Scala philosophorum*, the 12 steps of his ladder which is an image of that of the patriarch Jacob. It is the central sun in which the divine spirit of wisdom has put his tabernacle. It is the spiritual Bible closed with 7 seals or clasps [from] which, being once opened, incorruptible light will presently shine forth. And to this the true Alchemists allude the 7 days work of creation, the period of which is the sabbath or day of rest, the suns day, also their 7 inhibitions, or 7 distillations . . .

It is therefore the mystical Book of the Law, which being closed and locked up brings death, but being opened . . . proffers life . . . says St Paul, the letter kills but the spirit vivifies, and Moses said to the people in the delivery of the tables in which the law was written: I call God to witness that this day I have proposed life and death to you, *choose that you may live*. This book therefore at the first was a transparent and spiritual stone as formed and graven with the finger of God, but through the rebellious, idolatrous and stiff-necked disposition of the Israelites it was broke, and an obscure stony one was afterward presented to them in lieu of a clear and transparent one, so that after the time that the bright table was broke, Moses had a veil before his face, and even as his veil was over his brightness so was his speech a cover over the truth of Gods mystery.

It was also that very stone on which Jacob laid his head [in margin: Genesis. 27] when he slept and saw the vision, of which he said when he awaked: Surely the Lord was in this place and I was not aware. How fearful is this place. This is non other but the house of God, and this is the gate of heaven. And he took the stone that he had laid under his head and set it up as a pillar and poured oil on it, and he called the name of the place Bethel, which is as much to say as the house of God. Then he concludes thus: And this stone (said he) which I have set us, shall be Gods house. It follows therefore, if the house of God then is consequently the mystical church, the spouse of Christ with him is her soul, of which Solomon makes so large testimony in his song, who called herself the rose of the field and the lily of the valley, after the pattern whereof the Alchemists have shaped their red and white Elixir, or stone.

This stone therefore of Jacob does well agree with the celestial Jerusalem which is as it were the body, house or city of the spiritual Lamb, which serves to govern and illuminate it, even as the mental beam of the soul of man is directed for the guide to his body. So also was Jacobs stone the house or body whose lord and soul was God, for he calls it the house and dwelling-place of God ... It is therefore the white stone mentioned in the Apocalypse ... It is the spiritual rock which fed the Israelites with meat and drink, that real and material rock (I say) that Moses struck with his rod in the wilderness ...

By this therefore I demonstrate that the Elixir (which amongst the Chemists has his denomination from perfection of maturity as the Hebrews Thummim of the perfection of light) is a material, yea and an earthly substance but of an exalted nature by a supreme purification ... for when there is a mutation of elements by a complete rotation of them, wherein earth will be turned into water and water into invisible air, and it into fire, then does fire conclude all with a spiritual, celestial and a bright golden earth which is the tabernacle wherein the light of wisdom, so much mentioned of the wise king [in margin: Sap: 19:17], abounds, the which is full of life and multiplication ...

But now I have told you what this spiritual body, or temple or rock is and have proved it to be material, I will proceed briefly to the incorruptible and bright spirit or soul which dwells within it, by the act and virtue whereof it is exalted to a metaphysical immortality and plusquamperfection by which all things multiply both in the excellency as well of quantity and body as in quality and virtue of spirit, for the body being exalted draws his like to it magnetically, as the salt of Tartre (which was air at the first condensed into salt) sucks the crude air to it and multiplies his quantity, which before it was freed from the body of the Tartre and mundified from his impurities it could not do . . .

This inward and central brightness is the golden beam of formal perfection, the glittering seed of the true and simple philosophical and Theological light, which is sowed in the clear and transparent virgin earth or vessel of purity. It is the simple emanation without any respect to the creature, issuing from the fountain of light. It is the essential or formal centre and circumference, the beginning and the end, the all in all, the life and uncreated light of the world, the middle and central soul of the sun, the life or soul of the elements, the incorruptible breath, spirit and being of all things, which are composed or shaped by it, the queller of confusion, the just spirit of reformation, the head, author and act of multiplicative propagation. It is the pagans god, their Demogorgon, their Jupiter, their powerful Cupid or God of love, their bright Apollo, the internal original from whence their wise men and learned priests have under types, hieroglyphic and enigmatical allegories derived the descent of their gods, which the blind and ignorant miscreant has made the foundation of his Idolatry by not understanding the spiritual meaning and sense of the wise men, and thereupon has taken and worshipped the creatures in which this divine light dwelled instead of the light itself which is in the centre of every soul . . . It is the Jews Great God Jehova, Elohim, Adonay, their Hochma or wisdom. It is the Christians Messiah, their assured word of God, their light of the world, their spiritual Christ, their Jesus, their Lord and saviour from whom and by whom they expect salvation both of body and soul. It is the Cabalists *Aleph lucidum*, or bright Aleph, their first emanation from the fountain of light or bright unity, their second numeration of Sephiroth called Hochma, their Angel Metattron, the true sabbath or mark of rest. It is the natural philosophers central and formal being or beginning, their *Actus primus*, or first Act, or Agent, their *natura naturans & infinita*, or creating and infinite Nature, the moralists *summum bonum a quo omnia bona*, their mark or goal of felicity, the end of all material appetite, the metaphysics *Ens, unum* and *bonum*, the true Alchemists formal stone of plusquamperfection, the soul (I say) of their material Elixir before described, the central and inaccessible brightness of their spiritual gold, at whose aspect all imperfection is exiled and banished, yea and the very Prince of Darkness himself (. . . not being able to endure the brightness thereof) is constrained to vanish from the presence hereof. The mystical Hebrews term it their Urim, or light dwelling in Thummim which is perfection . . . It is the physicians truest *aurum potabile*, their perfect *elixir vitae*, and most certain antidote to cure all venomous diseases . . . It is wisdom which is the infallible medicine which as far surpasses the inventions of Hippocrates or Galen as light in beauty does excel darkness. It is the Grammarians Alpha and Omega, soul of the spiritual harmony and congruity of his speech, the Rhetoricians eloquence and grace, the Arithmetitians unity in simplicity which is the beginning of all number, and the Hebrews make it their final or punctual Aleph which signifies 1, and their great Aleph which imports a 1,000, beyond which Aristotle affirms that there is no denomination . . . so that we see that Aleph signifies both Alpha and Omega, the beginning and the end, and therefore also it is their trinity in perfection, for 3 seems to return from the binary confusion into the unity from which it came, and therefore 3 is the root of the progression of all formal perfection: it is their number from which all numeration issues. It is the Musicians unison in itself, which is the fountain from which all other symphonical accords flow, as each number from unity; and it is their Diapson which is all one with the philosophers wished end or *summum bonum*, for this is the chord or consonant of the highest perfection in music, and completest proportion which operates in the wide spirit of the world, by the harmony of which all things are exalted to the highest pitch of perfection . . . and the mean harmony by which the binary number lives and subsists in this world . . . it is the accord of the body in composition . . .and the body is varied according to the inconstant tune . . . and thus is the whole fabric of every thing in this world framed and tuned with a greater or lesser accord in perfection according to the ordinance of the unison who multiplies more or less the degrees of perfection in his creatures . . .

Now concerning man himself, which is by reason of his perfection called *omnis creatura*: it is his mental beam, by the presence whereof he is exalted above all other creatures in reason and understanding . . . He breathed into Adam the spirit of life and made man after His own image, and ye are the temple of the Holy Ghost . . . it is everywhere; for Solomon says the spirit of the Lord fills the earth . . . To conclude, it is all in all and singularly in everything . . . But experience has taught the sagest philosophers that this excellent and most bright spirit is to be found more plentifully in some things than in others, and therefore they have made choice of that thing which is most rich and plentiful in the presence of this benign and bright spirit, to cause his occult light to shine out of the darkness thereof; and yet nevertheless this light is so sacred and divine that it will not appear but to such as are worthy of her presence; neither shall they find it or have a real view of her in her spiritual palace, or shall observe her to appear out of her dark cloud except (as our saviour teaches us) they seek that they may find, neither shall it be opened to them except they knock . . .

Lo therefore it is evident that a manual operation and mental action are co-operators in this divine magistry and therefore Art is most necessary in the adapting of kind unto kind, which is the principal exploit in this business. But the true operation of this mystery differs so far from the Vulcanian and torrid artifice of spurious Alchemists as white is from black or light from darkness. It is a blessing in our art (which ought to be slow and humble) that we crave of Him who is the only giver of wisdom, for . . . man proposes (as the old phrase runs) but God disposes . . .

For art does but prepare and separate the pure from the impure by the adaptation of things that are not out of kind. As for example we must prepare and cleanse our hearts, thereby to make them fit vessels to receive these blessings of wisdom, not by a moral discipline, but by a spiritual refinement, for wisdom, which is pure and simple, must have the like purity to welcome her and not [be] imaginary and supposed, nor that which is not what it appears vulgarly to be; for foreign and strange fire is not acceptable to God (as it appeared by the error of Aarons sons for which they paid dearly): it is the fume of Gods fire, it is the brightness of his fiery gold whose odours are acceptable in the nostrils of wisdom, and to the act of none other sacrifice will she vouchsafe to incline . . .

I conclude therefore this point, and say that as this is a divine act or form of the highest degree of exaltation, so it requires a matter to work on like unto itself in proportion, the which matter therefore ought to be a spiritual and dignified Earth made fit within and without to receive so divine an influence, for this incorruptible spirit will not covet a lodging that is not worthy of his presence, wherefore this spiritual or heavenly earth must be a pure virgin and a clear transparent vessel in which it deigns to reveal itself to men . . . I could bring forth an infinite company of . . . testimonies to demonstrate that the Philosophers intention was to produce a material and that to such an end they prepare and refine a certain elementary or passive matter to such a perfection that it might be worthy to receive so excellent and divine a form or act as the beam of wisdom is whose tabernacle is a spiritual or aereal fire . . .

But I will show you here what is the power of that hidden wisdom which inhabits, as is said, every creature more or less, and by how much the more she abounds in any of them, by so much the more noble are they and dignified before God, for wisdom delights to inhabit clean vessels and pure hearts . . . St Au[gu]stin, in his comment upon Genesis, By how much the more (says he) the creature abounds in light, by so much the more does he participate of divinity. And Divine Plato with Apuleius his precept was: Speak not of God without light. It was the breath of God that did dignify man only (amongst other creatures) with the mental beam of reason and understanding whereby he got the pre-eminence over all the creatures of the earth. It was the multiplicative excellency and benignity of refreshment and nourishment that made our

Saviour to choose bread or wheat, to allude to His precious body. It is the perfection of God amongst the minerals that Holy Scriptures themselves hold in so high esteem by reason of His perfection and exaltation in mineral brightness and light ... And all of these perfections proceed from those grains of the first formal excellency, which God alloted to gold in the creation thereof.

Man therefore being the prince of the animal kingdom, having a supernatural spirit alloted to him, brandishing, though occultly ... divine fire, has the gift, if rightly he understand himself, to abound in spiritual generation and propagation, which is chaste and holy before God and man ... And this multiplication is a pattern to that of the Elixir of the Philosophers, or the Elixir the type of it when it is multiplied in form or essential quality of light and not in quantity of substance. We may observe that man generates and propagates after the manner and condition of beasts; but this kind of multiplication is not spiritual, but is corporeally to the bestial gross carnal or outward man, as the chaste and spiritual begetting issues from the inward man.

Wheat also in the vegetable kingdom multiplies in itself, but this kind of multiplication ... is after death by a due resurrection ... The bands of the 4 elements are so strongly fastened in the original composition of wheat before a natural corruption, that the bright spark of vegetative multiplication ... is kept in, and cannot otherwise dilate itself, but only persists to maintain the grain and appurtenances thereof in the true shape and existence of their first creation ... The small portion of light or life in the grain inhabiting the centre thereof, being by putrefaction and dying set at liberty, draws to it a multitude of invisible and formal fire of his own likeness in essence which lurk invisibly in the air, after that they are descended by influence from heaven, to the end that by the assistance and aid of their bright wings it might with the greater clarity sublime out of the dead and dark putrefied body, that being freed from the slime of the dissolved mass of elements, it might fly up to his native country, the kingdom of light and perfection, namely into heaven which is the natural home of each quintessential substance.

But lo, in her ascent she is so beloved of the element of fire; which by a natural contiguity & concatenation challenges the next place to the aethereal spirit, that she cannot ascend without this element, and the like consanguinity challenges the element of air [with] that of the fire, and again the water [with] that of the air, and lastly the earth, keeping the selfsame proportion and order with the water as it had with the air, not being able by reason of his natural inclination and ponderosity to climb any higher, holds and ties down the other aspiring and soaring elements, which already in their voyage towards heaven have attained to the top of the lofty stem or shaft of the wheat, which she confirms with fast roots within her bosom.

Hence therefore it comes that the spark or formal grain of wheat, having in the little space of his liberty drawn many times his own quantity of his like nature out of the worlds universal spirit (operating after the pattern of this saying, When I am exalted I draw all spirits unto me); and being glued round about by the subtle slime of the 4 inferior elements, which serve to multiply in quantitative substance or matter as the fiery seed doth in quality, the specific shape makes it to impart a share of his abundance to other parcels of the thin and covetous elements, which concur to a manifold formation of many other like grains according to the nature and habit of the vegetable spirit which is the vessel of wheat. Thus much we may safely speak as having proved it by an ocular demonstration . . . Thus therefore have we expressed briefly the chief and original cause or central act of the multiplication of wheat, which is the principalest of vegetables and all other inferior creatures of his kind, that we may with the better apprehension of the reader conceive the secret manner of a double multiplication in the prince of minerals, which is Gold.

It is not unknown to the true Philosophers that not only gold has his fix[ed] and bright grain in him, but also all other metals; nevertheless it chances that in some of them this grains action or motion to the complement of perfection is hindered by the impurity of an heterogeneal sulphur which is joined to his pure mercury, and for this cause is lead called *aurum leprosum*, a leprous gold, for although his appetite to perfection be complete, yet is the impurity of his sulphur a main bar or let in his progression to his desired mark . . . Roger Bacon in his *Mirror of Alchemy* [in margin: *Specul. Alchemiae* c. 3] admonishes us to make election of such a matter as has in it a pure, clear, white and red *argentum vivum* or quick silver which is not as yet produced to perfection, but remains in his natural composition equally and proportionably mingled, that is, by just measure with pure sulphur of his like nature, and congealed into a solid mass, that by our wit and discretion, and by the help of our artificial fire, we may reduce him to his complete purity and make him such a thing as after his complement may be a thousand times stronger and more perfect than those simple bodies that are only concocted by natural heat.

Hebraeus . . . says: The stone or medicine that you seek pullulates and multiplies as the grain of wheat after it is mortified in a good soil. He concludes there also that the only mover or agent in Nature is a vegetable spirit which is the proper fire of Nature, which is hot and moist and therefore the Philosophers term the Elixir their vegetable stone. *Novum lumen* instructs us at large that minerals have their seeds as well as animals and vegetables whereby they multiply when their gross matter is separated from them and all impure obstacles are taken away . . . but . . . [the] matter ought to be reduced by Natures fire to the highest perfection of a mineral nature before it can work out his effect of transmutation in the mineral kingdom . . .

I conclude therefore this long journey, which truths golden harrow has so faithfully traced through this troublesome furrow, with this assertion: that here is a material Elixir spiritually exalted even as our bodies shall be after perfect resurrection, which is the palace wherein the incorruptible spirit or bright and beauteous essence of form dwells, at which the hieroglyphics, morals, mysteries, parables, allegories and oracles both of sacred scriptures and divine philosophy chiefly aims at.

Lastly we observe that vegetables are transmuted into the substance of the animal, namely, into flesh and blood, and also we are ocular witnesses that serpents, oyster shells, whales bones and other such like animal parts have been changed into hard and rocky stones, as in the King of France his cabinet in the Louvre at Paris and [in] Huntingtonshire I have seen. Also it is commonly known that [the] wood and roots of trees have suffered the like mineral transformation. It follows reciprocally that minerals may be turned into vegetables, and they again into animals, and therefore minerals may have the selfsame gift by reduction to multiply (as Artefius witnesses in his book called *Clavis major Sapientiae*) [as] that [of] animals or vegetables.

The Eleventh Furrow

[Scot asserts that it was against the philosophers' doctrine to conceal their knowledge.]

[Fludd:] It is the glory of God to hide the thing [in margin: Prov. 12]; and in another place, the subtle man hides his science. And Christ admonishes us that we be as subtle as [the] serpent. And he would not that pearls be cast before swine, and to that end he commanded his disciples to speak in allegories; Mercurius Trismegistus calls his silence the fruitful propagation of goodness; Plato says, that those things which are sacred in mysteries are not to be divulged; Pythagoras and Porphyrus did consecrate their disciples with religious silence. Orpheus did exact an oath, with a certain terrible authority of religion, of them which were candidates in the ceremony of the divine mysteries to keep silence lest the secrets of religion should come to profane ears ... Apuleius in his mystical secrets says I would declare it if it were lawful ... also Paul saw things which are not to be revealed. In the great Rosary of the Philosophers we read that ... [he is] in danger of death, or an Apoplexy, which reveals this mystery unworthily. Picus Mirand [in margin: In Proem: lib. 2] witnesses that the custom of the ancients was in the writing of great things either physical or divine, to do them hiddenly and figuratively, because the rudeness of the hearers which could not endure the splendour of Moses his doctrine, ought to have doctrine and words with a veiled face, lest such as were to be illuminated should be blinded with too much light . . .

Notes

1. Excerpted from C. H. Josten, 'Truth's Golden Harrow: An Unpublished Alchemical Treatise of Robert Fludd in the Bodleian Library' in *Ambix* 3, Nos. 3 & 4 (April 1949), 91–152.
2. The bracketed words appear in the manuscript.

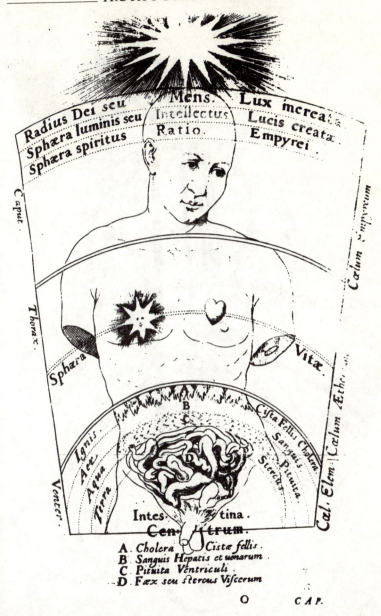

Figure 5: The Microcosm–Macrocosm Correspondence related to the Three Heavens.

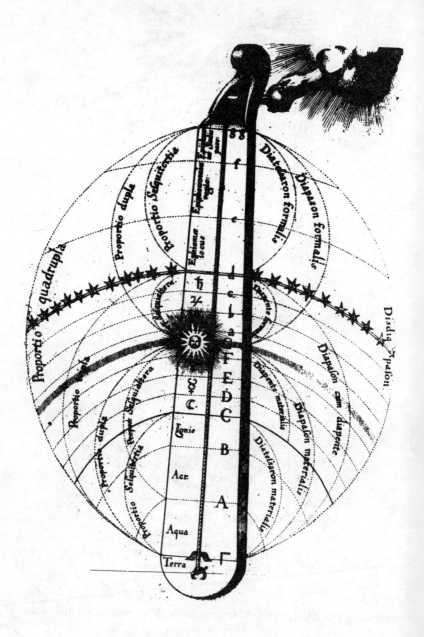

Figure 6: The Celestial Monochord.

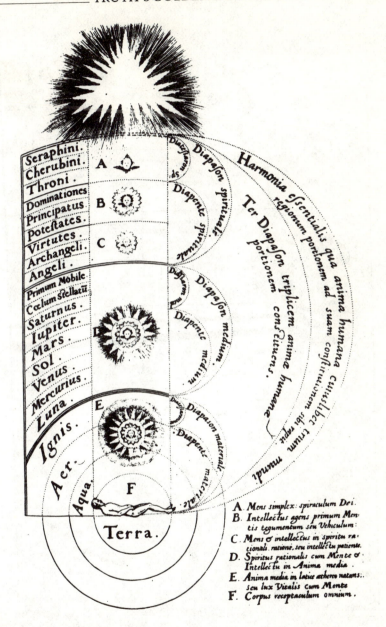

Figure 7: The Harmony of the Human Soul with the Universe.

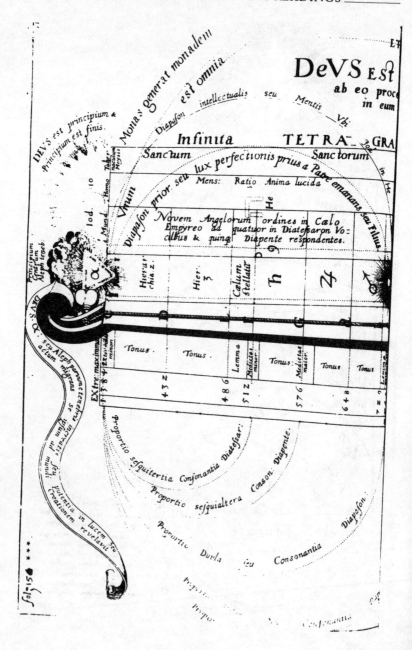

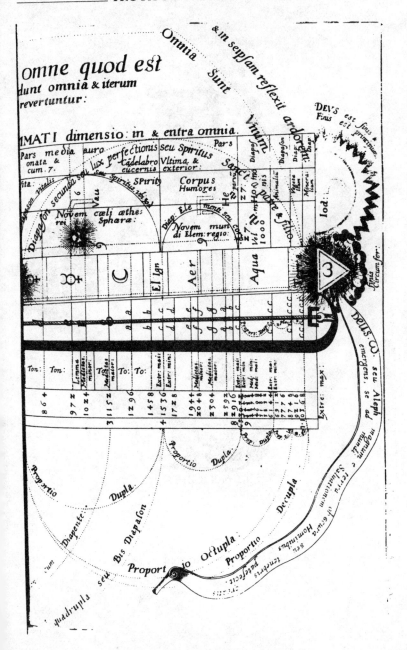

Figure 8: The Universal Monochord.

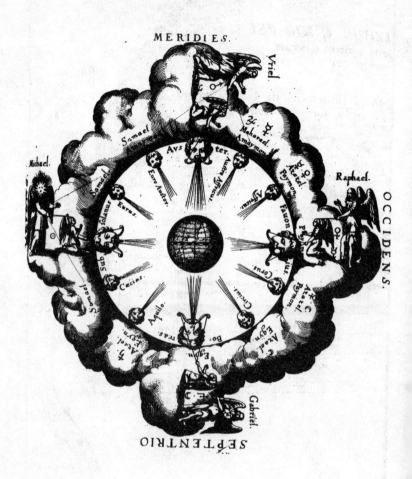

Figure 9: The Winds directed by Angels.

Doctor FLUDDS Answer unto M. Foster
OR,
THE SQUEESING OF
Parson FOSTERS SPONGE, ordained
by him for the wiping away of the
WEAPON-SALVE

Wherein the Sponge-bearers immodest carriage and behaviour towards his bretheren is detected; the bitter flames of his slanderous reports are by the sharp vinegar of Truth corrected and quite extinguished; and lastly, the virtuous validity of his Sponge, in wiping away of the weapon-salve, is crushed out and clean abolished.

PSAL. 92.7
Bilis acutissima aceto correcta acerrimo redditur dulcior.
Opera Dei, vir brutus & stultus non intelligit.

The Assertion of Parson Foster and his Faction or Cabal is this:
The wonderful manner of healing by the weapon-salve is Diabolical or effected only by the invention and power of the Devil;
But, the royal Psalmist [is] guided by the spirit of God, saith Psal. 71.18.

Blessed be the Lord God of Israel, who only worketh wonders: Therefore, the Prophet pointeth thus, at these and such like enemies of the Truth. Esa. 5.20. Woe unto them that speak good of evil, and evil of good; which put darkness for light, and light for darkness; that put bitter for sweet & sweet for bitter; woe unto them that are wise in their own eyes, & prudent in their own sight.

LONDON
Printed for Nathanael Butter, 1631

The Contents of this Treatise

1. Taketh away and utterly disannuleth those scandalous reports which Master Foster hath in his writing most falsely and irreligiously divulged and laid unto the world, how unseemly a thing it is, for a man of his calling to accuse and censure his brother unjustly.

One, doth answer particularly unto every objection, that Master Foster doth make in a generality for the abolishing of the weapon-salves usage.

This small 2. Is divided treatise is into 3 parts divided into or Chapters: 3 Members, of the which whereof the the

Other, doth maintain Theologically the Cure of the weapon-salve, to be good and lawful, & proveth it by the authority of holy writ, to be the gift of God, and not of the Devil.

Lastly, demonstrateth the mystery of the weapon-salves cure, by a Theophilosophical discourse, and showeth how it is grafted or planted by God in the Treasury of Nature.

Last, doth answer unto each particular objection, which our Spongy Adversary maketh against a certain Treatise, expressed by me in my mystical Anatomy, for the proving and maintaining of the cure by the weapon-salve to be natural, and in no way Cacomagical.

To the well-minded and unpartial Reader

I did not think (Courteous and learned Reader) to have stirred up the puddle of this mine Adversarys turbulent Spirit, for a 3-fold reason, whereof the first is his insufficiency to undergo a task of so high a nature and so far beyond his reach or capacity, namely: to dive into this profound Mystery of curing by the weapon-salve; and then because my learned friends, observing his inclination unto immorality and malice (as appearing indeed more skillful to Cavil and calumniate, than to decide with gratuity so weighty a controversy) gave me counsel to have patience and to answer such a manner of man with Silence; and lastly, by reason of that reverence which I bear unto his vocation, namely, as he is a minister of Gods word, professing unto the world that I would endure much rather than to have the least opposition with any of that profession. But since I have perceived his indiscreet importunity to extend itself so far as to urge me beyond the bounds of patience by setting up in the nighttime two of the frontispieces or Titles of his Book, as a Challenge, one each post of my door, and understanding of his many other indecent actions, as well by hearsay, as in his public writing, wherein he hath in a scornful and opprobrious manner laid disgraceful matters unto my charge, I have been forced, against my will, to take the person (setting with my best respect aside the Parson and his habit) into my better consideration, and to examine in this my small Pamphlet the misdemeanours of his Book entitled *The Sponge to wipe away the weapon-salve* expressed both in his erroneous doctrine touching the main subject of that manner of curing, as also his rude and unseemly carriage towards his bretheren, that thereby I may as well clear myself from such immodest and unjust imputations, which with an evil conscience he hath laid upon me, as also expresseth shallowness of the person in this business, which he so vaingloriously hath undertaken.

I doubt not but as there are many who in every degree can discern an evident difference between this satirical gentleman & myself, so because I know that there are diverse of the common sort of people, who in their zeal without understanding are apter to conceive and judge amiss, than in their charity to ponder the truth of the business, I am the willinger to cope with this unsavoury philosopher, whom I find (and I make no doubt, shall prove) fuller of windy verbosity than of solid philosophy, or any thing else that is grounded upon firm reason.

What shall I say unto the man, whom, unto my best remembrance, I never saw nor knew, save only by a bragging smoke of rumour, which pronounced me . . . [from] far-off an anathema? The thunder which long smothered in the gloomy cloud of report is now broke forth; the flame of his lightning assaileth me. What then is more convenient and requisite than sharp vinegar to quench it? Yea, he shall find it so acute & piercing (though not with railing and calumniating edge, according unto his bitter custom, but reserving itself within the bounds of Christian modesty), that his Sponge shall not be able to drink it up or wipe it away. It shall quell the insatiable appetite of his salve-devouring Sponge, and squeeze or crush it so, that it shall be constrained to vomit up again that wholesome child of Nature and gentle friend unto mankind (I mean the weapon-salve) which it hath drunk or sucked up, and leave it in its wonted splendour and reputation amongst men.

And lastly, it shall examine the quintessence of the Sponge-bearers self-conceited wit, and tell him, that what sometimes appeareth great, is not always the same it seemeth, but rather a shadow or blast of empty air. This is all (judicious Reader) that I will say at this time, as for the rest, I refer it to the proof in the pondering whereof I most heartily pray you that all partiality or peculiar affection being laid aside, you will be pleased faithfully to judge of this our controversy, and weigh every passage thereof in the just and equal balance of your best discretion.

Your Servant in a greater matter,
Rob. Fludd

THE FIRST MEMBER

CHAPTER 1

Esteem it no point of indiscretion in me, first to abolish and take away all such reproachful imputations as have been wrongfully laid to my charge, that with the greater courage unto myself, and better acceptance and satisfaction unto my countrymen, I may proceed unto the main business or question which is proposed by this mine adversary.

Unto him therefore I must in the first place turn the edge of my pen and file of speech, for as much as he, forgetting that I am his Brother in Christ, and his Country-man, yea, and not differing from him in Religion, should so far neglect the precepts of the Prophet David, our Saviour Christ, his Spiritual Master, and his Apostolical followers, as slanderously and void of Christian modesty (most requisite unto a person of his divine calling) to publish unto the world (although it were deservedly, much less against the grounds of Truth and rules of Justice) the weakness and imperfection of his Brother.

The Kingly David therefore saith: *Thou givest thy mouth to evill, and with thy tongue thou forgest deceit, thou speakest against thy brother, and objectest slander unto him, therefore art thou in darkness because thou hatest thy brother* . . . also the Apostle James . . . maketh a difference between that wisdom, which is from above, and that which is from beneath, in these words: *If ye have bitter envying and strife in your hearts, rejoice not: neither by liars against the Truth, for this wisdom descendeth not from above, but is earthly, sensual and devillish; for where envying and strife is, there is sedition and all manner of evil works. But the wisdom that is from above, is first pure, then peaceable, gentle, easy to be entreated, not judging, and without hypocrisy, etc* . . .

Thus far I have spoken with the mouth of the Prophets, our Saviour, and his Apostles. But mine adversary will reply and say, that Doctor Fludd is a Magician, and hath maintained a damnable and Diabolical action, namely, the curing by the weapon-salve, to be good and lawful, and therefore by warrant of Scripture he ought sharply to be told of it, etc.

I answer, That from this objection may arise a double question; whereof the first is, Whether the assertion of our adversary be true or false? and then, whether it be a brotherly part, first to divulge it, though it were true indeed, unto the ears of the people, before he hath admonished his brother in private of his error.

Touching . . . the first, whether . . . I am a Magician or no, it shall be fully discussed in the third Chapter . . . where I . . . prove to each well-minded person (and that to the burden of my rash accusers conscience, if he have any) that he is justly ranked and numbered amongst those, at whom the Prophet David, our Saviour Christ, and the Apostle James have aimed in the places above mention[ed].

As for the second, our Lord Jesus resolveth it in these words: *If thy brother hath trespassed against thee, go and tell him his fault between thee and him alone* . . . whereby it is apparent, that mine adversary hath not dealt with me as one Christian ought to deal with another, forasmuch as he hath violated the precepts of his Master Jesus Christ in this point, and therefore this his misbehaviour can no way be excused . . .

Chapter II

My reputation doth a little engage me to answer certain extravagant passages, which our Sponge-bearing author maketh against me; as for example:

Doctor Fludd (saith he) hath writ in the defence of the weapon-salve, well he may, he is called, by Franciscus Lanovius, Medico-miles, a soldier-Physician, and being a weapon-bearing Doctor, may well teach the weapon-curing medicine, especially setting the Armiger before the Doctor, the Bun before the Gown, the Pike before the Pen.

(Good Reader) observe this unreasonable jest; rather than fit out, the gentleman will pick straws or play with a feather. What? Not one word with reason, nor yet any syllable in good rhyme, but all upon the letters G and P? An excellent argument of a simple wit . . .

He would seem to teach me that the Armiger or Esquire ought to be set before the Doctor, being that there is a question, whether a Knight or a Doctor should challenge the first place, and therefore (saith he) the Esquire was ill placed before the Doctor, being that the Doctor is the better man; verily I think that it ought to be so, and yet nevertheless, it is certain that nowadays a reasonable Esquire thinketh much to yield place unto a Doctor. But this is not to our question. He seems to accuse me because I say, *Robertus Fluddus Armiger & in Medicina Doctor: Robert Fludd Esquire and Doctor of Physic* . . .

And now I will express the cause, why I put the Esquire before the Doctor. It is for two considerations: first, because I was an Esquire and gave arms before I was a Doctor, and being a Knights son; next because, though a Doctor addeth gentility to the person, who by descent is ignoble; yet it is the opinion of most men, and especially of Heralds, that a gentleman of antiquity is to be preferred before any one of the first head or degree. And verily for mine own part I had rather be without any degree in University than lose the honour [that] was left me by my ancestors. Thus far I have thought fit to satisfy the gentleman in his humour, wondering at nothing more than that he should leave the main matter to snarl at my gentility. Would he think it decent in me to revile him for his lowness of birth or ignobility? For I know what he is [but] God forbid such an absurdity should come from my pen, much less to upbraid him with his gentility, if he were a gentleman indeed.

In another place he inveigheth bitterly against me in the manner:

The Doctor, who impiously attributeth composition unto God, dareth to attribute corporality unto Devils; the contrary of which, that they have no manner of bodies, is the tenet of the Church.

I see that our Master in Arts is scarce in this matter his Arts master: he talks with *Mersennus* the Friar his tongue, and therefore is but *Mersennus* his parrot; wherefore as I have answered him in Latin, so will I partly in the same sense satisfy this importunate author in English, not with impious terms, according to his custom, but modestly . . .

[Fludd quotes from Wisdom 12.1 and Job 10.9.]

To conclude: I have answered this point more at large in

the Reply I made to *Gassendus* his Retection of my Philosophy, where I prove that the virtue of God is in every thing, as it composeth all things . . .

CHAPTER III

What therefore can I expect from his outrageous pen, but slanders of Witchcraft, Magic, and such like abominations? . . . His scandalous texts are these:

Doctor Fludd hath had the same censure passed on him, and hath been writ against for a Magician, and I suppose this to be one cause, why he hath printed his Book beyond the seas. Our University and Bishops are more cautious (God be thanked) than to allow the printing of magical Books here.

Surely D. Fludds very defence of the weapon-salve is enough to make it suspected, himself being accused for a Magician by Marinus Mersennus, with a wonder that King James (of blessed memory) would suffer such a man to live and write in his kingdom . . .

In the first place . . . for the better satisfaction of my Countrymen and friends, I will express the cause, why I touched the superstitious *Magia* or Magic of the ancient Ethnics. My scope was and hath been to write as well the natural discourse of the great world and little world, which we call Man, as also to touch by way of an Encyclophy or Epitome of all Arts, as well lawful, which I did commend, as those which are esteemed unlawful, which I did utterly condemn, as superstitious and of little or no probability at all; among the rest where I came to speak of the Arts, which belong unto the little world or Man, I mention the Science of Genethlialogy, which treateth of the Judgement of Nativities, wherein I produce the great dispute which did arise between the two famous Philosophers *Porphyry* and *Iamblicus*, whereof the first did hold that a man might come to the knowledge of his own Genius or good Angel by the Art of Astrology, namely, by finding out the Planet and Nature of his spirit, that was Lord of the eleventh house, the which by the Astrologians was for that cause called *Bonus Dæmon* or the good Angel. But *Iamblicus* his opinion was, that a man had need of the

Assistance and Knowledge of a higher Spirit, than was any of those which were Governors of Fatality, namely, of such Intelligences as were ascribed to the rule and direction of the seven Planets; wherefore I did thereupon express the superstition of the Ancients with the Impossibility thereof, that thereby I might the better decry it & make it the more ridiculous to wise men . . .

Now judge (all ye that are unpartial and truly learned) what an offence was here to decide according unto my power, that great Controversy of these two notable and eminent Philosophers, which hath stuck and been undecided even unto this day, being that I in the conclusion ascribed the whole Glory unto that sole and only Spirit, which is the Prince and Lord of Angels and Spirits. I professed to write generally of all, but as I went along, I distinguished the good from the evil, that men might the better beware of, and refuse the one, and make choice of the other. Now therefore, that I have expressed unto you the ground why this our Critic and his Cynick Master the Friar *Mersennus* have slandered me with the Title of a Magician, I will proceed to the answer of every member of his frivolous objections.

Doctor Fludds defence of the weapon-salve is enough to make it suspected.
      And why I pray you?
      Marry because he himself is accused for a Magician by Marinus Mersennus.
      The conclusion is much like the capacity of the concluder: Doctor Fludd is suspected for a Magician *ergo* the purge of Rubarb which he prescribeth, or rather any point in Philosophy or Physic, by him maintained in writings, is Magical. *Non sequitur argumentum* . . .
      M. Foster, if his eyes had been so favourable and his will so charitable as to have looked on my answer to *Mersennus* in the defence of that point, before he had judged, he would without doubt, as well as hundreds of other men, whereof some are Churchmen, of no mean rank, and many Doctors of Physic of excellent learning, have averred, that D. Fludd had

answered *Mersennus* so fully, as well in that accusation, as all other points laid by him unto his charge, that he could not be able any way to reply against it. And it is well known here in England to such as have been conversant beyond the Seas, that the sufficiency of my answer hath so satisfied the learned, as well in Germany as in France, that he hath been by them much condemned for his slanderous writing, and esteemed of but meanly for his small learning and indiscretion.

Now the ground of his malice unto me was, for that he having written of the Harmony of the World, and finding that a Book of that subject set out by me was very acceptable to this Countrymen, he invented this slander against me and my Harmony, that thereby he might bring his own into the better reputation. But what did I say? That he was accused and condemned for that his slander by some in France, yea, verily his dearest companion, who by reason of his insufficiency, was easily persuaded to take his cause in hand, and to answer for him, I mean *Peter Gassendus*, his friend and Champion, chideth he *Mersennus* for such his uncivil and scandalous reports against me in these very words:

And although (my Mersennus) the zeal wherewith you are moved against Fludd is to be commended, nevertheless you cannot be ignorant, how grievous and intolerable a thing it is unto any man that liveth in the Christian world, to be called a Witch, or evil Magician, a Heretic-Magician, or a teacher or divulger of foul and horrible Magic, and that such a teacher is not to be suffered unpunished, also to provoke the King or Prince to punish him, and besides all this threaten him; saying, that for that cause he should be drowned or drenched in the eternal Lake and so forth.

Besides the Atheism and Heresy, which also you object and lay to Fludds charge, verily, these are things which would stir up the patience of Ruffinus or S. Jerome, whereof the one when he requireth patience in other things, yet he concludeth, that he which can bear or dissemble with the offence of one heresy, he crieth that such a man is no Christian. The other saith, I will not that any man should be patient in the suspicion of Heresy, much less to be accused or suspected of Atheism or naughty Magic.

In which words our English world may discern first, how

this my honest-dealing and moral foreign adversary doth check his uncivil friend, whose part he undergoeth, for his immodesty and small discretion; and next doth teach my home-bred adversary a great deal of manners or behaviour in writing against an adversary, namely, not to contend with foul and scandalous language, but with acute arguments, and those to the purpose, armed with the truest reason of Philosophy.

Doth not Master Foster blush now to see his mighty and magnanimous author *Marinus Mersennus* checked by his judicious friend, whom he himself hath elected as well for Umpire and Stickler in his cause? Yea, and a chief Champion to defend it, for calling me unjustly a Magician and other misbeseeming names? Yea, is he not ashamed, if he hath any [shame], to choose out a lying and false author for the propagating of his Brothers slander? If this be not so, ask *Gassendus* . . .

To which, for his better satisfaction, I produce this other place out of *Gassendus* his Reply against me, in his friend *Mersennus* his behalf:

To the first Chapter of the third Book, wherein Fludd is cleared from the suspicion of Atheism, Heresy, and especially of Devilish Magic

[Gassendi is speaking to Mersenne:]

There resteth now the suspicion of Evil-magic, of which especially the question is made, but this is an evident argument unto me, that he is no such Magician, because he doth not believe (or at leastwise maketh semblance not to believe) that there are such Devils as we imagine to be familiar with Witches.

Let Master *Foster* therefore see, upon what fickle foundation he hath laid the false and malignant slander of an Evil-Magician on me. But alas! I smell a Rat (for I will use his own witty phrase): he careth not how he may disgrace anyone, so that he might thereby the better serve his own turn. Because I have produced in my Mystical Anatomy a natural reason for the weapon-salve, which he neither can, nor, for all his poor reasons expressed in his Book, shall be able to resell; therefore, forsooth, I must be numbered amongst the Magicians.

And wherefore? Marry because *Mersennus* hath given the same censure on me. And who is *Mersennus*? A railing Satyrical

Babler, not able to make a reply in his own defence, and therefore being put to a *Non plus*, he went like a second Job in his greatest vexation to ask counsel of the learned Doctors in Paris. And at last for all that, he, fearing his cause, and finding himself insufficient, procured by much intreaty his friend *Peter Gassendus* to help him, and called another of his friends unto his assistance, namely, one Doctor *Lanovius*, a seminary Priest, as immoral as himself, and one that professeth in his Judiciary Letter much, but performeth little.

And in good faith, I may boldly say, that for three roaring, bragging and fresh-water Pseudophilosophers, I cannot parallel any in Europe, that are so like of a condition, as are *Mersennus*, *Lanovius*, and *Foster*: all three exceeding terrible in their bumbasting words, imagining to quell and make subject unto their thundering braves the stoutest Scholars of Europe, if they cared for them, and did esteem them more than Bugbears to scare away crows or frighten little children.

As for *Peter Gassendus*, I find him a good Philosopher, and an honest and well conditioned Gentleman, just as well unto his adversary as friend, not passing beyond the bounds of Christian modesty, but striking home with his Philosophical arguments, when he seeth his occasion . . .

Marinus Mersennus (saith Master Foster) doth wonder, that King James of blessed memory would suffer such a man to live and write in his Kingdom.

To this I answer that King James of everlasting memory for his Justice, Piety, and great Learning, was by some Envious persons moved against me, touching the same subject; but when I came unto him, and he in his great wisdom had examined the truth and circumstance of every point, touching this scandalous report, which irregularly and untruly was related of me, he found me so clear in my answer, and I him so regally learned and gracious in himself, and so excellent and subtle in his inquisitive objections, as well touching other points as this, that instead of a check (I thank my God) I had much grace and honour from him, and received from that time forward many gracious favours of him. And I found him my just and kingly Patron all the days of his life.

And must I now after so regal a Judge have such an upstart Inquisitor as is Master Foster, to Judge and censure me again, and that by the ridiculous authority of an ignorant Friar, whose friend doth justly condemn him, and check him for his slanders, and clear me from all such crimes as he objecteth against Me?

Then he make a very shrewd objection, saying, *Because Mersennus writ against Doctor Fludd for a Magician, therefore I suppose that this is the cause, why he hath printed his Books beyond the Seas, our Universities and reverend Bishops, etc.*

Though I need not answer, in this point, a man of so envious a condition, yet, for charitys sake, which bids me not offend my brother, I will at this time satisfy him. I sent them beyond the Seas, because our home-born Printers demanded of me five hundred pounds to print the first volume, and find the cuts in copper; but beyond the Seas it was printed at no cost of mine, and that as I would wish. And I had 16 copies sent me over with 40 pounds in gold, as an unexpected gratuity for it. How now, Master Foster, have I not made you a lawful answer?

As for the University: I wonder my works should seem so mal-gracious unto it, when they are registered in two of her libraries. And surely, if my conscience had persuaded me that there had been anything in them which had been so heinous or displeasant, either to the Kings Majesty or the Reverend Bishops, I would not have presumed to have made first our late King James of blessed memory, and next three of the Reverend Bishops of the Land, the Patrons of them; being that I electing them my Patrons, [I] must present them with the first fruits, and therefore must know, that if anything had happened amiss in them, it could not be hidden from them, whom in verity I would be afraid to displease, as being such as with my heart I reverence.

[Foster:] *His friend Joachimus Frizius (or rather his own self, as faith Lanovius, in a Book called* Summum Bonum*) excuseth Roger Bacon, Trithemius, Cornelius Agrippa, Marsilius Ficinus, & Fratres Roseæ Crucis from being Cacomagicians, I wonder at nothing more than that Belzebub was not in the number.*

A singular Diabolical conceit!

For the first, whether that Book be mine or no, I have satisfied *Gassendus*, whose only ape *Lanovius* is. For he objecteth nothing, but what he taketh out of *Gassendus* his Book . . . only this much I will say for *Joachimus Frizius*, that what he hath produced out of their own works, in their own defence, excuseth them, and accuseth such calumniators as *Master Foster* is, who are so apt to condemn a person for that they are altogether ignorant in. Let the Readers observe the proofs in *Frizius* his Book to clear them; and then if any will afterward accuse them, I shall deem them partial.

But we must note by the way, that our Sponge-bearer must make election of Jesuits (as in his Epistle he confesseth) and Friars and Seminary Priests to be his instructors and teachers, to reply both against the weapon-salve and me, when he knoweth that they are such as can afford neither him nor me (as being esteemed among them for Heretics) one good word. Nay, I will tell him for his greater shame, that their only spite unto me is, because they discern my works to be well esteemed abroad in the world, my self being (as exorbitant unto their Church) esteemed by them an Heretic. All that the greatest adversary I have, even *Marinus Mersennus* himself aimeth at, is to have me change my Religion, & to gain me to their side, & for that intent he promiseth me, if I will leave my Heresy (as he termeth it), many rewards & courtesies. But I find here at home even amongst our own Religion, some men less friendly and greater enemies unto me and mine honest endeavours that abroad.

Mersennus his words are these, after he had thought with great terms to terrify me:

But if you Robert Fludd will leave your Heresy, I with my friend will heartily embrace you, and will either face to face speak with you, or by letters confer with you about certain sciences, and I will desire him not to write against you; but that you may be received by the Grace of the divine power amongst the children of the Catholic Church, that you together with us may eternally celebrate the divine praises in the place of bliss; if not, thou wilt be tormented with eternal flames, as it is certain that Heretics shall, and those that go from the Catholic

Religion, which your ancestors did embrace: especially such, as persist obstinately in their Heresy, will certainly be damned. For Gods Word is true and infallible, wherefore examine seriously your conscience.

In another place he wisheth, that leaving my Heresy, I would join with them in the correcting of Arts, telling me, what an applause I should have for so doing, of every Commonwealth.

This I speak to some of my Countrymens shame, who instead of encouraging me in my labours (as by Letters from many out of Polonia, Suevia [Swabia], Prussia, Germany, Transylvania, France and Italy I have been) do prosecute me with malice & ill speeches which some learned Germans hearing of, remember me in their letter of this our Saviour Christ his speech: *No man is a Prophet in his own Country.* It was not for nought the wise man said: *He that addeth unto himself Science, contracteth unto himself much pain and vexation* . . .

THE SECOND MEMBER

CHAPTER I

The main scope of the whole business is contained in this question, which he proposeth thus:

Whether the curing of wounds by the weapon-salve be Witchcraft, and unlawful to be used? . . .

His main arguments in his first Article:

All lawfull Medicines produce their effect either by Divine Institution, as Naamans washing himself in the River of Jordan to cure his leprosy [or] the Pool of Bethesdas curing such as entered in after the Angels stirring it; or else by natural Operation, according to such virtues as God in the Creation indued such Creatures with, whereof the same Medicines are composed, as the lump of figs to cure the impostume of the King Ezekiah, as the Wine and Oil, with the which the wounded man was cured by the Samaritan.

But this weapon-salve worketh none of these ways ergo the Cures done by it are not lawful, but prestigious, Magical and

Diabolical.

The minor is denied . . .

Unto your first reason I answer, that it doth not follow, that because it is nowhere registered in Scripture, therefore it is not of divine Institution: what, because figs, wine and oil, yea, and clay tempered with spittle, are noted in Scripture, for external Medicines, therefore must the use of caustic, veficatory, healing, fluxing, and such like other external Medicines daily used by Christian Physicians, be reputed for unlawful Magical & Diabolical, because they are not registered in Scripture? Or is nothing instituted by God, but what Scripture maketh mention of: how then can that saying of the Apostle be true, that *God worketh all and in all?* If all and in all, then worketh he also all Acts and operations, as well occult and mystical, as those which are manifest and apparent unto sense: and therefore all Acts are instituted by God . . .

Unto your second I say, that it is of no more validity than the first. The main axiom of the vulgar Philosophers, upon which you ground your proof for the excluding of this salve out of the list of Nature is this: *Nullum agens agit ad distans.* Upon this you frame out this argument.

[Foster:] *Whatsoever worketh Naturally, worketh by corporal or virtual contact; But this worketh by neither:* Ergo *it worketh not naturally.*

First, concerning that axiom in Philosophy, I know and can prove it by experience to be false. For the fire heateth *ad distans.* The lightning out of the cloud blasteth *ad distans* . . . The Sun and fire do act in illuminating *ad distans.* The Loadstone doth operate upon the Iron *ad distans.* The plague, dysentery, small pocks, infect *ad distans* . . . nevertheless . . . I affirm, and it is evident to every mans capacity, that this medicine doth cure by a virtual contact, namely, by a sympathetical property, which doth operate *inter terminum à quo & illum ad quem*, between the beginning and end magnetically and occultly or mystically . . .

Again, we deny that it is the artificial composition made with mans hands that cureth, but the natural ingredients of the composition, which God hath originally endued with such an occult and mystical virtue in curing . . .

Lo, how he [Foster] accordeth with his great enemy, that damned Magician *Paracelsus* (as he termeth him) who affirmed that it was *Donum Dei*. As concerning his authors which he citeth against it, I esteem them not: there are as many for it of a better authority and judgement. For they are neither your Schoolmen, who deal only in imaginary speculative Philosophy; nor *Joannes Roberti* the Jesuit, and such like fantastical Theorics; but learned Physicians, great Philosophers, both theoretically and practically profound in the mysteries of Nature, and therefore the fitter persons to discuss a business of this physical nature.

Amongst the which I nominate, in the first place, the Bishop Anselm, who for his integrity, deep learning and holiness of life, is canonized a Saint, and then amongst the deep Philosophers and Physicians, which have been conversant in the Mysteries of God and Nature, *Theophrastus Paracelsus*, who terms in justly *Donum Dei, Cardanus, Joannes Baptista Porta, Oswaldus Collius, Joannes Baptista ab Helmont*, and many other excellent and well experimented Philosophers and Physicians, who as well by the practical art of Alchemy (than which there is no Science in the world that doth more ocularly bewray and discover the hidden mysteries of Nature) as other assidual observations grounded upon proof, and not on imaginary contemplation only, have like true philosophers, dived into this mystery of healing: men (I say) who have been as subtle to eschew, and wary to foresee the Devils craft, yea, and to distinguish his act from the of God in Nature . . . And although some superstitious Physicians of this kingdom (such men as are apter rashly to judge this business than to ponder it with due consideration) may seem to be adverse unto it, yet, they cannot choose but know, that there are many thousand things more that are hidden in the secret closet of Nature, than commonly man doth know; or can at the first discern . . .

CHAPTER II

And therefore if God operateth all in all, then the Devil

operateth nothing; but curing is an operation, and therefore a work only of God . . .

And this operation of God, as well by himself, as in his created organs, doth extend itself; but also unto [the] arcane or hidden; yea, and to such as are miraculous & wonderful, even as this cure by the weapon-salve appeareth to be unto the fantasies of worldly men, making them to admire and wonder at it, as a company of birds do at an owl in an ivy bush, censuring after the wisdom of this world diversely, and that according to every mans imagination. Some boldly and presumptuously proclaim . . . [it] to be the work of the Devil; some aver it to be a main foppery and vain imagination in too credulous persons who by having only a good opinion of the thing, are cured; some condemn it, as a superstitious and abominable manner of healing, for as much as the election of ingredients must be done by an Astrological observation; and others, approaching nearer the truth, term it a Natural *Magia*, or a Magnetical or secret act of Nature; and some more essentially grounded, and religiously observing the prescribed order of holy writ, do (as true Christians are bound to do) refer both this miraculous and wondrous act in curing, & every other wondrous work besides, unto that glorious God, who hath made both heaven and earth, as assigned to them by his spirit, as well those virtues which work in the eyes of wordlings miraculously or wonderfully, as others which appear more familiar unto their sense . . .

And consequently not any Devil; nor Angel; nor man; nor medicine; but God only performeth it: by that spiritual gift of healing, he hath imparted unto man and his creatures in their creation and continued it in them from generation to generation . . .

CHAPTER III

. . . we are taught that the life, form, and nature of every Creature doth essentially spring and proceed from God, and therefore what gift of healing is found to proceed either from compounded or simple medicines, be they Angelical,

Celestial, Elemental, or of an Animal, Vegetable, or Mineral composition, it proceedeth from their Creator, as being either bestowed upon them in their Creation for that wholesome purpose, or else miraculously and beyond the common course of Nature imparted unto some Creatures to effect. And therefore man ought not rashly to condemn a medicine, because it worketh after another manner than the vulgar doth, for God hath alloted unto some medicines, occult & hidden properties; and therefore work they not by an external and evident elementary quality . . .

In like manner can no man express any natural reason that is manifest for the attraction of the Iron by the Loadstone, of Straw by Amber, or why the Loadstone looketh towards the North, or why the Laurel or Bay tree preserveth from the harms of lightning and thunder, and likewise how directly this cure is effected, etc. The causes of these things are occult and hidden unto the common philosopher . . .

It is apparent, then, that the incorruptible Spirit is in all things, but most abundantly (next unto the great world) in the little world called man: for as in the great world, God is said rightly by *Jerome* his translations (leaving the corruption of others) *to have put his Tabernacle in the Sun*, from whence by a perpetual and never dying motion, he sendeth forth life and multiplication to every member and creature of the great world, and by the agility of his Spirit [see Figure 5, p.171] . . . he moveth and giveth life unto the whole Spirit of the world . . . the spirit in the Sun animateth and giveth motion, life and spirit unto the aerial spirit of the whole world . . .

So also, and in the very like manner, the same incorruptible spirit filleth the little world (*it is the Temple of the Holy Ghost*) and hath put his Tabernacle in the heart of man, in which it moveth, as in this proper macrocosmical Sun in Dystole, and Diastole, namely, by contraction and dilation without ceasing, and sendeth his beams of life over all the whole frame of man, to illuminate, give life, and circular motion unto his spirit . . . also as this abstruse spirit doth give heat by his activity and essential motion unto the great world, the very same it doth effect in the little world, and all things else, when it doth not

. . . rest or withdraw his own act within itself . . .

As this Principal and central mover in the spirit of each world doth radically and solely act and move essentially in and over all, namely, from the centre to the circumference, his *Primum mobile*, or first moved in the great world, is the principal Ætherial region or sphere, by the circumrotation whereof, the Sun (which as David saith, is a vessel full of the Glory of God) is wafted about the earth in 24 hours, that thereby the whole spirit of the world may be recreated with life, vegetation, and multiplication.

And therefore this Spirits first and most worthy sphere, in which it centrally doth move, is the Quintessence or Ætherial spirit of life, which by his presence is vivified and animated; and this Ætherial spirit being the immediate vehicle of that incorruptible spirit of life, is carried in the grosser elementary or sublunary air, by which medium it penetrateth, into animal, vegetable and mineral bodies by inspiration or expiration in animals partly occult, as by the pores of the body, & partly manifest, as by the lungs, & in vegetables and minerals occultly, and only to be perceived with intellectual eyes, and so giveth life & multiplication to every thing . . .

To conclude: as to create, vivify, and sustain each creature, he put on all things, so he saith: *Spiritus incorruptibilis in est omnibus*; and again, *Spiritus Dei implet omnia*: whereby it is evident, that this divine and incorruptible spirit, by which we live, move, and have our being, is in man, for without it he is dead, a snuff, a nothing: his place therefore, or the heaven wherein it moveth, is our ether, or heavenly spirit, which acteth invisibly in our aerial vehicle: the grosser and coarser part whereof is blood, as well vital or arterial, as natural and venal.

Hence came those especial ordinances, or legal precepts, which were given by God, touching the blood, not only of man, but also of beast . . . because the blood is the seat of the Soul or vital Spirit, which is inspired by God . . . Here therefore observe, that the Spirit of life is from God, who vivifieth all things: the life of the blood and fat is in this spirit and of this Spirit, wherefore it is written in another place, *You shall by no means eat the blood and the fat, for the life of the flesh is in the blood* [marginal note: Lev. 3; Acts 17.24]. It is easily therefore to be discerned, what a concatenation here is of members in succession, which derive their lives from one and the same radical essence or spirit, and are made by it to sympathize with one and the same harmony in the creatures composition; being that He hath made of one blood all mankind . . . and consequently being all flesh and bones are made of one blood, there must be a great relation between them and mans blood in general, and consequently between the blood and the ointment which is made of them.

These things therefore being rightly pondered, as infallible grounds, we conclude thus: in the question proposed, we are to observe these five objects: namely, first the wound, secondly, the blood which issueth from the wound, thirdly, the manner of conveyance from the ointment, to the wound at any reasonable distance, fourthly, the nature of the ointment, and lastly, the manner of operation, whereby the cure is effected.

First, therefore concerning the wound, it is a violation of the work, which the spirit of life did effect, namely, an effusion of blood; in which the spirit of life is carried and moveth, a hindrance and diverting of the course of the natural humours, a division and solution of the fat, flesh and other such like parts from their integrity and continuity, an offence unto that peaceable act of life, effected by that incorruptible spirit of God, which by this His property, or attribute, is apt to vivify all in all. For, this cause therefore is this radical, acting spirit interested in this business, or unnatural action; as finding his work hindered, and his essential action disturbed by the wound or violence offered. For, whereas the blood is the vehicle of it, and his vivifying act was to circulate in the organical blood, and to cause transmutation of it into flesh, and other parts for vegetation & multiplications sake, and for the preservation of the *individuum*; now is the same blood sluiced out at the mouth of the wound, and made inutil and of none effect: the body (for the animating of the which this secret spirit is ever diligently inclined) is debilitated and made drooping. Wherefore as the incorruptible and vivifying Nature hath intended to rectify his human spirit by her lively activity, so verily is she ready to oppose all violence offered, and to correct & repair again, all that which violent irruption hath caused . . .

Secondly, the blood, as it is the vehicle of the spirit of life, though it be by the wound voided out; yet retaineth in it this spirit of life; but in another property: for, it doth not now act to live, that is to say, it doth not send forth his beam from the centre to the circumference to cause life; but contrarily, being as it were displeased with the violence of the act, contracteth itself from the circumference into the centre, that is, from action in the circumference of the creature, into itself, being contracted into the centre thereof, where it seemeth to rest, and so leaveth his bodily and aery vehicle as congealed, stupefied and dead . . . yea, this spirit doth entirely leave, and forsake the flesh of the dead, being that his life (as it is said) is in the blood . . .

In the third place, I come to the manner of conveying of the blood from the wound unto the ointment. The blood is taken from the living fountain of blood in the wounded, either as it is smeared on the weapon that did the deed, or as it is fastened on some stick, iron or other thing, and so conveyed unto the ointment at any reasonable distance. Now a reason is to be showed, how it is possible that there can be any certain relation between the wound and the ointment: for (as Mr Foster saith) there may be castles, hills, walls, and gross air between the ointment and the wounded, which may hinder the cure.

First, we must remember that we have expressed in our precedent discourse the excellency of the animating spirit, in whom is all the virtue and each property of the four winds, and being [as] it is this spirit of spirits . . . what (I pray you) can hinder his act or operation? And with what distance can his activity be limited, being that it is the spirit of the winds, and the soul of the lightnings, and the essence of the Sun and stars of heaven, which by his animation do cast their beams peripherally unto every angle of the horizon, or hemisphere?

Can this spirit, because contained in mans blood, not penetrate many hundred miles by emanation out of his bloody vehicle ... Whereby it appeareth, that his very spirit, by which man breathes, cannot be limited in his penetrating and extensive dimension; nor yet hindered in his passage by any intermediate obstacle.

To conclude, that man, that believeth and relyeth on this spirit, may effect what he desireth. For, even by the true knowledge and use of it, the Prophets and Apostles did wonders, as well in curing as effecting matters of greater admiration. This spirit therefore, which is called intellectual, as he maketh to understand ... vital in respect of his vivification ... and natural in respect of vegetation and multiplication ... doth act and shine forth by secret beams, unto that object of the dead blood, which is carried from it unto the ointment, in which amputated blood lurketh a portion of spirit, resting without action.

Now the nature of the one is rejoiced in the nature of the other, forasmuch as both do sympathize together, being that they are all of one consonance or degree, or unison in vital love: as for example, I take two lutes or viols, or any other such like instruments, I set one of them at one end of the table, & set the other at the other end, I put a small straw upon one of the strings, of the one lute, which importeth A-la-mi-re, or De-la-sol-re, and then strike the Gam-ut of the other lute, and the straw will not once stir, because they do not sympathize in one sound and proportion of waving air, therefore have they not a relation one unto another: so also, if the blood be carried unto any ointment heterogeneal in nature unto the party wounded, it will do nothing in this cure; but if you put a straw on the Gam-ut, or A-re, of the one, and strike the other on Gam-ut, or A-re, being unisons, you shall perceive, the straw presently to leap of the other string, by reason of the over great vibration of loving activity, and like for all proportion, which the sympathetical harmony, betwixt each strings, causeth to other in the air, yea, this effect will happen, though there be put boards or other such like obstacles as may hinder the direct line of the vibration in the air or medium betwixt the two lutes ...

For as both are but one spirit, though they seem to differ in distance, as do the chords of both lutes, so likewise are those two tones but one tone . . . and therefore make but one unison. But because the one spirit cannot essentially be separated from the other, no more than can divinity essentially be divided into parts, as also the one tone cannot be essentially distinguished from the other . . .

The fourth to be considered is the ointment and his nature. Who, but a mere idiot can deny, that like doth desire his like, or that one nature being stronger, doth cherish, foster and relieve another that is weaker, and the weaker rejoiceth in the aid and comfort it bringeth? The ancient Physicians and Philosophers have observed that lungs nourish lungs, and brains nourish brains that are weak, the spleen helps to fortify the spleen, & for weak guts we make glysters of boiled guts, the stomach of a cock helpeth digestion, the very spittle voided by the physical lungs [is] said to cure the lungs, worms mortified and dried to powder, destroy worms . . . Now if we look into the composition of this medicine, we shall find, that it is of a wonderful consonance with the blood of man . . . Therefore without doubt, there is the selfsame relation of unison betwixt this ointment with the blood in it and the wounded mans nature . . . And for this cause will be apt to evibrate & quiver forth one mutual consent of sympathetical harmony, in that the spirits of both, by the virtual contact of one anothers nature, be made by conveying the individual spirit of the one into the body of the other, that the lively balsamic virtue of the one may comfort and stir up the dull and deadly languishment of the other . . .

. . . because the ointment is made of mans blood, mans fat, mans flesh, or mummy, and the fumous excrescence of mans bones, called *uznie* or the moss that groweth on the skull, according unto my receit, and for that the nature of the catholic spirit thus specified is in the ointment, though not working, and is stirred up to operate by the union which it hath now from the beams of the lively and operating spirit of the wounded . . . Therefore the mixtion of these two spirits now operating, in one vivifying union, makes them to tend

unto the fountain of life, as the grain rising out of the earth
. . . But because this earth or salve is more spiritual, it tendeth
out his power unto the blood, by that harmony, which the
continuation of spirit doth effect, namely, as it were by an
unison, by reason of the uniformity of the specific spirit
belonging unto man, by the union whereof the four discordant
elements, and every member of mans body are united unto a
sympathetical harmony, adopted to the use of life in the
creature, yea, so forasmuch as the blood, flesh, fat and bones
in all other unreasonable creatures are framed out of one kind
of elementary form, and fashioned alike, by the same operating
spirit; it is no marvel if his blood, being brought unto the same
ointment, do[es] also cause health in the wounded creature,
being it doth generally tend to life, which is proper to ally
bloody creatures, no exceptions had unto each specific
difference . . .

CHAPTER IV

. . . we see that by the application of salves, balsamics or
inward physic, the natural spirits and internal actor of life doth
help and assist the medicines in their cure, or else they would
not effect any such matter. For this reason is the Physician
called *Audiutor naturæ*, the helper of Nature.

Now, that the principal agent of this cure is comprehended
in the body of the wounded, I prove it in this manner: it hath
been averred, and will be maintained by persons of great
knowledge, not babies, but of a far greater maturity, both in
learning and judgement, than our Sponge-carrier, that in their
manifold experience they have observed, and found it most
true, that when the wounded person hath sent his blood on
a stick, iron or weapon to the place of the ointment, and that
thereupon he hath been in a good way of healing; if in the time
of his cure, he hath to do with a menstruous woman,
immediately the curative power in the ointment is lost, and it
will do him no good; also they have oftentime observed that
if the wounded person happen in the mean season to have an
issue of blood out of his nose, the curative property will be

quite taken away . . .

There is another admirable experiment tried by a noble personage, of whom I will make mention more . . . for one of his men having deeply cut his finger, and that about the joint, with a scythe, as he was mowing of grass; his finger bled still, and could not easily be staunched. Wherefore the Earl wished them to knock off the scythe from the handle, and to bring it unto him that he might anoint it; the wounded fellow went about it himself, and at the very first knock that he ha[d of] the weapon that had wounded him, the blood staunched, and he bled not one drop more . . .

CHAPTER V

There is a Knight dwelling in Kent, a man judicious, religious and learned, called *Sr Nicholas Gilbourne*, one (I say) with whom I both am, and have been, long familiar. For he married my Sister. This Knight having good acquaintance with one Captain *Stiles*, forasmuch as in time past he was his tenant, was with the said Captain in the Company of very good and learned Divines, at the making of the said Ointment, who saw all the ingrediences apart, and after beheld an Apothecary to compound them together without any kind of superstitious action, where it was generally adjudged to be a lawful medicine, and no way superstitious or Diabolical. A box of this ointment was bestowed on this my brother-in-law; what wholesome effects it hath wrought, I will in a word relate unto you, and that verbatim as I have it under his own hand.

The first (saith he) was at Chat[h]am in Kent, where the servant of one *Poppee*, a shipwright, was cut with his axe into the instep, so deep as it could pass, and not cut it off; upon the hurt (which was in the afternoon) he was brought unto me; but I refused to meddle with it, only I advised him, to wash his wound with his own urine, which he did. The next morning early I did dress the axe, and after dressing it, I did send to know, how the fellow did? Answer was made that he had been in great pain all the night; but now lately was at ease. The next morning coming into my study, I struck my Rapier down upon the axe, the hilt whereof struck the ointment off from the axe, which when I found, I sent to understand how he did? and had answer, that he had been exceeding well that night; but this morning he was in great pain, and so continued. I therefore anointed the axe again, and then sent again unto him, and heard that he was then at great ease; and within seven days was perfectly well . . .

His second history of this manner of curing was this: I (saith *Sr. Nicholas Gilbourne*) having given unto me the sowing of a pond at Charing; after I had done, the boys of the town went into it to seek for fish. Among the rest *Brent Deering* (the son of Master *Finch Deering*) did go into it for that purpose, and there had a reed run into the calf of his leg. This bled much, & put him to great pain, which caused his mother to send unto one *John Hart* a surgeon of Charing to search and dress the wound; but he continued after dressing full of pain, and apt to faint. Whereupon his sister was sent unto me, to do my best for his ease. I answered that I could do no good, because he was already dressed [by] the surgeon. But that would not satisfy them, and therefore upon their importunity, I advised them (because they informed me that the orifice was very narrow) to wash away all the surgeons work, and to put a knitting needle into the wound, so far as it would go, and to tie a thread where it would pass no further. They did so, and found that it went quite through to the very skin of the other side. This knitting needle I did anoint, and in four or five days, [the boy's leg] was well in appearance, saving that upon the top of the orifice, there was a dry scale. I was nothing well satisfied, to find that it was not perfectly well, but had still a scale remaining upon it. And therefore I did newly anoint the knitting needle overnight, and the next morrow, there came out of the orifice a small splinter of the reed, and after that, it was in two or three days perfectly whole . . .

By Windsor, there was one, who had somewhat to do in the Chase or Forest, who, as he was mowing of a piece of meadow, fell backward upon the edge of his scythe, and cut all his back so dangerously, that his life was greatly to be feared; the scythe forthwith [was] sent to London to Captain

Stiles, who . . . anointed it, wrapped it up, and laid it aside. Not long after, there came one to demand for Doctor *Stiles*, and he was sent unto Do. *Stiles*. The Minister, who, understanding that it was about a thanksgiving for a cure done by the weapon-salve, sent him unto the Captain, [saying] he desires to speak with him. The Captain, being at dinner, or supper, with diverse of his friends, sent for the fellow into the dining room, and there he told the Captain, that the wounded man did acknowledge of him his life under God, assuring him that the dangerous wound did heal apace, after he had sent his scythe unto him, and was thoroughly cured without any other application. And for a part of recompense, and to express his thankfulness, he presented him with a side of Red Deer.

Here you may see, that his cure was performed at a distance of 20 miles between the wound and the ointment.

Chapter VI

. . . I think it not amiss to certify each Reader, that there is a certain noble Personage of this kingdom, very religious, judicious and learned, who at the first scoffed at this kind of cure, as a thing impossible, and after that he perceived that it was true indeed, he was terrified by such scarecrows as Mr *Foster* is, to put it in practice, forasmuch as he was made believe, that there was a prestigious deceit or Cacomagical virtue and operation in it. For which cause he did abstain from the use and practice thereof; although he did acknowledge that act of it to be wonderful. And yet nevertheless, because his curiosity did incite him, to dive a little further into the truth of this mystery, he did at the last desire to speak with one Captain *Stiles* (a man well known by his acquaintance, to be both wise and religious, as also adverse unto all superstitious actions or ceremonies) because he was noted to be a great practitioner in this manner of cure. His earnest desire of him was first to see with his own eyes, for his better satisfaction, every particular ingredient apart, which went into the composition of this salve, and afterwards to observe each ingredients preparation, and lastly to behold their mixtion or union in composition, that thereby he might the better discern, whether any unlawful act or Diabolical superstition did concur with the making thereof. The Captain consenteth, whereupon this Noble man, to be the better informed in this matter, called unto this composition a learned Divine, and a well-esteemed Doctor of Physic, who with one consent, after the complement of the business, did affirm, that there was neither any damnable superstition in the making of this ointment, as was falsely suggested . . . And for that reason, they did jointly conclude, that both the medicine, in itself, and the practice of this cure were natural and consequently lawful for any good Christian to use.

Hereupon this Honourable Personage, did for a twelve-months space, with happy and fortunate success, practice this manner of cure, on many that were wounded; and yet for all this, it should appear, that some busy Buzzards, or rather buzzing flies of this nature, did put into his head new suspicions, insinuating unto him, that the Captain might use some secret superstitious means, or unknown charms in the collecting, or preparing of the principal ingredients, which he could not discern, and that without this, those mystical effects could not be wrought, whereupon on twelve-month being past, he undertook for his more assurance, to make the composition himself, and to have the ingredients gotten and prepared by his own direction, namely, the moss of mans bones, etc. And for this cause he maketh Mr *Cooke*, the Apothecary, to beat into fine powder, such of the ingrediences as were to be powdered, and afterwards to compound [the weapon-salve] and to make it up, which when he had effected, he found, that this his own composition had the selfsame healing virtue, and prosperous success in curing that the other had; by reason whereof, he rests ever since, so confident in the safety and lawfulness of this cure, that not one of these fantastical Butterflies, by their painted shows without any solid substance, can alter his mind from this practice . . .

The above mentioned Noble Personage and Captain *Stiles*, with Sir *Bevis Thelwell* (who had his ointment from that Noble Personage, and hath performed by it many strange and

desperate cures) and Mr *Wells* of Dedford (a learned and honest
Gentleman) have cured (as they will make good) at the least
a thousand persons by this manner of cure, and now there are
many other, as well men as women, which have got of this
weapon-salve, and do daily an infinity of good in this
kingdom. From hence (I say) commeth the grief unto the
Surgeons, as well of this City of London, as of every Country
about. *Hinc dolor, hinc lacrimæ*: and have they not good reason
for it, when they lose such a mass of practice as would well
have stuffed their Pouches . . .

CHAPTER VII

Sir *Nicholas Gilbourne* relateth in his letter unto me these words:
The last time (saith he) the Lady *Ralegh* was at Eastwell at the
Countess of Winchelsey her house, we falling into some
discourse concerning the Sympathetical ointment, she told me
that her late husband, Sir *Walter Ralegh*, would suddenly stop
the bleeding of any person (albeit he were far and remote from
the party) if he had a handkirchers, or some other piece of linen
dipped in some of the blood of the party sent unto him. If this
were done by the Devil, I presume, that so wise a Personage
as was Sir *Walter Ralegh* would have left, or a least-wise not
have used that trade of curing or stopping of blood . . .

I was, whilst I did sojourn in Rome, acquainted with a very
learned and skillful personage, called Master *Gruter*, he was by
birth of Switzerland; and for his excellency in the Mathematic,
and in the Art of motions and inventions of Machines, he was
much esteemed by the Cardinal Saint *George*. This Gentleman
taught me the best of my skill in those practices, and amongst
the rest, he delivered this magnetical experiment unto me, as
a great secret, assuring me that it was tried in his Country,
upon many with good success. When (said he) any one hath
a withered and consumed member, as a dried arm, leg, foot,
or such like, which Physicians call an Atrophy of the limbs,
you must cut from that member, be it foot or arm, the nails,
hair or some part of the skin, then you must pierce a willow
tree with an Auger or wimble unto the pith, and afterward put

into the hole the pared nails and skin, and with a peg made of the same wood, you must stop it closed, observing that in this action the Moon be increasing, & the good Planets [be] in such a multiplying Sign, as is *Gemini*, and fortunate and powerful over *Saturn*, which is a great dryer. The selfsame effect (said he) you shall find in you, take the nails and hair, which is cut off the member, and close them in the root of an hazel tree, and shut up the hole, with the bark of the tree, and after cover it with the earth, and (said he) it hath been tried, that as the tree daily groweth and flourisheth, so also by little and little will the patient recover his health. But you must with diligence observe the motion of the heavenly bodies, and especially the places of the Sun and Moon, when this is effected. And to this intent, he did disclose unto me the time and seasons when the preparation unto such a cure should be effected.

But alas! What have I done? Now hath Mr *Foster* enough to cry out that this is Magic indeed; here is superstition in the highest degree. For did not he say, page 17, *that it is an Astrological, and therefore superstitious observation to collect any ingredient, or to do any thing by attending and expecting, when the Moon should be in such or such a house of heaven, and that by Scriptures, Astrologers, Magicians and Sorcerers, like birds of a feather are linked together?* His blindness leadeth him in this as in the rest.

For first he concludeth that all Magic in general is damnable and Diabolical, because one *species* or member of it is justly to be banished from Christian mens remembrance: as if there were not a natural Magic, by which *Solomon* did know all the mysteries in Nature, and the operations thereof; yea, as if the three wise Kings of the East did discover that the true King of the Jews was born by Diabolical Magic. How now Master *Foster*, were these three wise men Cacomagical Magicians, or such as the Scripture did allow of, and we Christians keep a holy day in their remembrance? Right, Friar *Mersennus* his ape! For he condemneth all *Magia* without exception of kinds, not remembering that *Magus* is in the Persian tongue interpreted a wise man or a priest. And in the very same manner, the Gentleman, after his Masters custom, condemneth all Astrology, for that members sake, which is truly superstitious and unlawful, not considering that the verity in both the true Magic and Astrology hath been falsely contaminated and abused by superstitious worldlings, and thereupon made the good, in the eyes of the ignorant, to be abolished and condemned with the bad, for the bad sake; and so goodness by vile men is swallowed, without any difference, by darkness.

I would therefore have our Sponge-bearing adversary know that there are four parts or kinds of Astrology in general. The first is conversant about the mutation of the air, and foretelling of tempests, diseases, famine, or plenty, etc. The second foretelleth the alterations of states, as also wars, or a pacific disposition in the minds of men. The third intreateth of the election of times, and of nativities. The last is directed unto the fabricating of characters, seals and images, the which, because it mingleth itself with superstitious actions, & is made an instrument for the abuses of impious persons; and especially because a Diabolical insinuation unto vice and impiety, may easily be perceived in it, is of all good Christians to be repudiated and condemned for unlawful.

What? Is the Almanac makers Science for this Mr *Fosters* exceptions to be put down, or must Physicians be forced to forsake or neglect their hours of election, in gathering of samples, or letting blood, or cutting the hair and nails, or stopping laxes, or making the belly lubric for this mans caveat? Doth not *Amicus medicorum* aver, *that the influence of heaven may help the working of medicines?* . . . and for this cause, *Haly* saith, *the Physician that is ignorant in Astrology is as a blind man, searching out his way without a staff, groping and reeling this way and that way.* And Ptolomeus, *that a good Astrologian may avert many effects of the stars which are to come.* Doth not *Galen & Hippocrates* speak much in their critical Treatises, of the necessity of observing the Moons motion? . . .

If this will not serve to stop our Adversarys violence, we will come unto the testimony of Scriptures for the confirmation as well of election of seasons, as to prove that the influence of the heavens doth operate as well good as evil effects. [Fludd quotes Eccl. 3; Eccl. 7; Ephes. 6 and Job 38.28.] For in this Text, we find the good influences of the stars are mentioned; and here also it is expressly noted, that the heavens have their powers on the earth.

I boldly affirm therefore, that all Astrology is not forbidden for as much as there is an especial observation to be had by wise men, of the influence of the stars. And for that purpose, there are hours of election, duly to be observed according unto this or that influence, which is most proper and convenient for our work. Again, whereas Mr *Foster* seemeth to make so slight account of the 12 Signs, and their essential operations on the earth; he may see, that such as have made their allateral notes upon the Text, do interpret the word *Mazzaroth*, to signify the 12 Signs, which do possess the 12 Houses of the Zodiac, which being so, mark the Texts conclusion: *Canst thou set the rule thereof on the earth?* whereby it is evident, that the 12 signs have an especial rule over the earth, and the creatures thereof, and that by Gods ordinance and appointment.

By this it is made manifest, that there is no Cacomagical superstition in observing times, days or hours, in which this or that star hath dominion, for the collecting of ingrediences, or preparation and adaption of medicines, or other matters, proper for the cure of man . . .

There is at this present, an honest religious Gentlewoman about London, that taketh an herb, called the Rose of the Sun, which hath small husks about it, which will open and shut, and she putteth it in plantain-water, and it shutteth and closeth up. She therefore, when a woman with child beginneth her labour, giveth her a little plantain-water, and though the labouring woman appeareth to the Midwife never so ready to be delivered; yet if the Gentlewoman see the vegetable closed, she concludeth that they are deceived, and that there is no such matter, and so it proveth indeed. Again, when the Midwife doubteth of her delivery, and yet she is indeed near it; the flower will open by little and little as the Matrix doth, and then the Gentlewoman bids the Midwife look to it, assuring her that she is ready for it, and it proveth so. This story was related very lately unto me by a Noble man of worth, and confirmed by a reverent Doctor and his Apothecary, who aver, that certain Midwives do at this day make use of this natural conclusion . . .

There is a Noble Personage in this kingdom of no mean descent, title and rank, among the English nobility, a most wise, grave, aged and religious Gentleman, I say, who hath cured a hundred in his time of the yellow Jaunders, the patient being 10, 20, 30, 40 (yea, & as he and others have reported) almost 100 miles off from him; and many of them, that he hath so cured, have lain long drooping under the burden of this disease, before they came to him, in so much that the use of common receits of Physicians could not overcome it. He hath both performed it by his servants at home, and hath communicated the secret unto some of his friends abroad, amongst whom he hath been pleased to rank myself. The urine therefore of the patient is sent unto this great Lord. His manner of cure is this: he taketh the ashes of a wood, commonly known and growing here amongst us in England; he maketh a paste of this wood with the urine, reserving a little of the urine apart for another purpose; this paste so molded & made up with urine is divided into 7 or 9 lumps or balls, and in the top of each of these, he maketh a small hole, and putteth in it a little of the urine remaining, & into those part of urine, he putteth a blade of saphron. And so without further doing, he puts the lumps in a secret place, where they must not be stirred lest the cure be hindered. And experience hath taught the world, many a score of ictertial men, or infected with the yellow Jaunders, have by this simple means been cured; and this is well known unto a 1,000 persons . . .

Now I will proceed unto the particular defence of mine own doctrine, expressed in my mystical Anatomy, against the which Master Foster doth inveigh so bitterly, and with so great a confidence. He crows there like a cock on his own dunghill, before he hath occasion, and challengeth gloriously the palm, and proclaimeth the trophy of his own praises, before he hath got the victory. The end crowns all, for truth is not bolstered up with high and bragging terms. It had been best for Master *Foster* to have heard me speak before he had publicly slandered me and set up the Titles of his Book on the posts of my door in my disgrace; whether it was discreetly done of him or not, I leave it to the censure of the world . . .

Each discreet Reader may discern by M. *Foster*s vehemency against me for composing in my mystical Anatomy the subject of this Member, that is more of envy and malice than for any defect in me, or offence committed by me . . . For out of doubt, he would not else upon so slight an occasion as was this short chapter expressed in the foresaid place, have so slandered me with the title of a Magician, as he hath; and alleged his wise Master the Friar *Mersennus* his authority for it, as profound an Author for railing and false slander as himself . . . Again, whereas this our home-bred Adversary saith, that I have excused myself from Magic in a Book intitled *Sophie cum Moria certamen*, and that *Lanovius* saith, *Cuius contrarim verum est*; for *Lanovius* (though in as malicious a manner as he could) doth clear me of that crime, alleging that mine unskilfulness or insufficiency in such things made him to think the contrary.

And therefore I must tell this my English calumniator, that there is a Star-chamber to punish such abuses, and consequently, he may perchance hear of me sooner than he doth expect, unless he bridleth his slanderous tongue the better hereafter. It is an argument of little Philosophy, and less Divinity, to rail unreasonably and scandalize with immorality . . .

The subject of this Chapter cited by me in my mystical Anatomy is only a discourse of the natural reference and Magnetic or attractive and sympathetical relation which is observed to be betwixt two distinct substances of the like nature, but differing in the distance of place: as between the Loadstone and the Iron; betwixt the blood and the salt of the same nature; in which the vegetating spirit, common unto them both, doth occultly abide. And you must note also (Courteous Reader) that in this particular Book of my mystical Anatomy, I did handle the secret and hidden properties of the spiritual or internal blood in the external, citing therewithall, as near as my small capacity would give me leave, the harmonical effects which it worketh, as well by contract or immediate touch, as at a distance.

I would fain know now, wherein I have offended in so doing? or how I have deserved M. *Fosters* slanderous ingression into his examination of this business? or whether in my natural discourse upon this subject I mention Diabolical Charms, Circles, Witchcraft, or unlawful and forbidden Characters, or such like? If you find nothing appertaining unto any such devilish Magic, then give your sentence, whether such a Prelude unto this business was honest, decent or anything appertaining unto the matter in handling.

As for the usage of the weapon-salve in itself, I protest before God and man, I never of my self did practise it unto this very day: but in my conscience, and by reason of a more strict inquiry, which for this cause I have made into it, I find it so free from any Diabolical superstition (which, God is my witness, I have ever hated, as I do the Devil and all his works) and have heard so much of the virtuous operation thereof, that from henceforward, malegree the demure writers or speakers against it, I will both practise it, and defend the lawfulness of it, as being more assured now than ever, that it is the blessed virtue of God; and not any act of the Devil, which operateth in it unto the health and allegement of Gods afflicted creatures . . .

And now for a farewell unto this my small Pamphlet, I would have my well-minded Countrymen to know, that, had not this rude and uncivil Adversary of mine, most untruly and disgracefully calumniated me, and laid without any just occasion unto my charge the unsufferable crime of Witchcraft or Magic, which is odious both to God and man, I would not thus far have hindered my greater business, and more weighty occasions to have satisfied his unreasonable and immodest appetite. And yet, I protest before God and the World, that I am so far from envying at his good qualities (if he have any), that in the first place, I pity his indiscretion and want of that modest and moral wit and behaviour in his writing which becommeth a true Philosopher . . . in the second place, I wish him with all mine heart more money in his purse, or else some good Benefice or Church-living to stop his mouth, the want whereof (as appeareth by his Epistle unto the Reader) maketh him in his writing first so forgetful of his Creators Omnipotency, that he presumptuously attributeth that, namely, the sovereign gift of healing, unto the Devil, which from all eternity belonged unto God; next, he most irreligiously and unjustly doth scandalize his Bretheren for ascribing that justly unto God, which only appertaineth unto Him; and consequently not to any Devil in Hell; and lastly, he seemeth to inveigh against some men of his own Profession, yea, and also to murmur against his Superiors in the Church, as you may collect, partly out of his Dedicatory Epistle, and partly out of that unto the Reader . . .

Let him therefore hereafter thunder forth, cry, & proclaim what he please (for such is his uncivil nature) I will from henceforth answer him (as a railing and cynic writer ought . . .) with Silence . . . But if he have some other business or subject that sticketh in his stomach against me (as I have heard he threatneth me with Mountains, and I am assured they will prove in the end but Mole-hills, as well as the preceding) perchance if I find him in his writing more modest and mannerly, as well becometh one that professeth the name of a Philosopher, & as a Master of Arts ought to behave himself towards a Doctor, who is his Superior; that is, if he strike hard and defend himself closely from being repaid with Theological and Philosophical arguments, and not with misbeseeming terms, foul-mouthed language, and false slanders, as his custom is; he will find that I will not refuse or fail him, but will be ready to cope with him in the Philosophical Camp of *Minerva*, when and how he dare . . .

Verbum Sapienti

The Final Summary

MOSAICALL

PHILOSOPHY

Grounded upon the

ESSENTIALL TRUTH

OR

ETERNAL SAPIENCE

Written first in *Latin*, and afterwards
thus rendered into English.
By ROBERT FLUDD, Esq;
&
Doctor of Physick.

LONDON,
Printed for *Humphrey Moseley*, at the *Princes
Armes* in St *Pauls* Church-yard. 1659.

To the Judicious and Discreet READER

My desire is (Judicious and Learned Reader) that I may not prove offensive unto any, if (in the imitation of my Physical and Theo-philosophical Patron St Luke) I mention and cite the testimony of Holy Writ, to prove and maintain the true and essential Philosophy, with the virtuous properties of that eternal Wisdom, which is the Foundation and Cornerstone, whereon it is grounded. Was not this the radical Subject of my foresaid Patron, who was as well a Divine Philosopher, as a Physician?

If the office of Jacobs ladder was for Souls and Angels to ascend from the Earth unto Heaven, and to descend from Heaven to Earth . . . if the chain of Nature has its highest and last link fastened to the foot of Jupiters chair in Heaven, as the lower is fixed on Earth, how is it possible for us earthly creatures, or rather divine Images, housed and obscured in clay tabernacles, to wade, of ourselves, through the confused Labyrinth of the creature to the bright Essence of the Creator; that is, to search out the mysteries of the true Wisdom in this world, and the creatures thereof, but by penetrating with a mental speculation and operative perfection into the earthly Circumference or mansion thereof, and so to dive, or attain by little and little to the heavenly Palace . . .

If God therefore in and by his Eternal Word or Divine Wisdom has first made the creatures, and sustained the same to this present; how can a real Philosopher enucleate the mysteries of the Creator in the creature, or judiciously behold or express the creature in the Creator (for in him are all things); but by such rules or directions as the only storehouse of Wisdom, namely the Holy Scriptures, have registered, and the finger of that Sacred Spirit indited for our instructions? Shall we with the Agarens, and those which were of Theman, forsake the Fountain of Virtue to search after true Wisdom where it is not to be found?

And yet nevertheless, lest my intention should by the misprison of any be ill interpreted or misunderstood, I think it convenient to certify you, that my purpose, in the progress of this Sacred or Mosaicall Philosophy, is far from any presumption to trench upon or derrogate from the deep and mystical Laws of Theology in her pure and simple essence, or to oppose the current of her Argument against those usual Tenets and Authentic rules in Divinity, which have been long since decreed and ordained by the Ancient Fathers of the Church.

But as it is certain that one and the selfsame place in Scriptures has a two-fold meaning, to wit, an internal or spiritual, and an external or literal; and either of these two senses are true and certain, though they seem to vary or differ by a diverse respect, not otherwise than under the name of one and the same man a double nature, namely a spiritual Soul and a material Body, are really to be understood. So also besides such mystical interpretation as the Texts of Scripture do internally contain; [they] may also express and delineate externally such created realities as belong to the true Subject of the most essential Philosophy.

And again . . . Theology points directly at the sincere and simple Nature with the virtuous extensions and powerful operation of the Divine Essence, making her demonstration *à priori* . . . Philosophy . . . moves by a clean opposite action or method, from the external of the creature or organ, *quasi demonstratione à posteriori*, to dive and search into its internal Centre . . . that is, to apprehend the Divine or eternal cause by the created or temporal effect . . .

. . . seeing that the Holy Bible does fully handle and set down the Subject of both these Sciences by the way of the two foresaid Demonstrations, namely as well after a Physical as Metaphysical manner, my hope is, that this my Philosophical Discourse will not be therefore sinisterly judged of by the truly wise and unpartial Reader because it chiefly relies on the axioms or testimonies of Scriptures . . . I could produce an infinity of . . . places out of Scriptures to manifest the universal acts and virtuous operations which are effected in the Elementary creatures by that most essential and eternal Wisdom, which is the main ground and true Cornerstone whereon the surest Mosaicall Philosophy does rely; but I esteem it needless, being that they are copiously expressed already by me in this my Philosophical Discourse, and therefore I imagine that these which are already produced will be sufficient to content and satisfy all such as are unpartially judicious; unto whose better wisdom and favourable constructions I recommend these mine endeavours, and finally, both them and myself, unto Gods blessed protection.

Your Friend
Robert Fludd

The First Book

CHAPTER I

. . . I purpose therefore with myself to make and forge me out an Armour of solid natural reason, and to temper it with the warrant of sacred authority. And lastly, I will make choice of ocular demonstration to serve me in this combat, instead of an unresistable weapon, or Herculean club to tame and subdue that unreasonable monster, *Incredulity*; than which there is no greater enemy unto mankind. And that I may the better accomplish . . . this design of mine, it is requisite . . . I ought to have . . . an experimental Instrument, or spiritual weapon, which may carve out a ready way to the truth by a manifest and infallible demonstration . . . I will make therefore election of such demonstrative Machines for my purpose as is vulgarly known amongst us whereby my intentions may be more easily understood of every man; and this Instrument is commonly styled by some, the Calendar-Glass, and by others, the Weather-Glass . . .

CHAPTER II

. . . the Instrument, commonly termed the Calendar, or Weather-Glass, has many counterfeit Masters or Patrons, in this our age, who because they have a little altered the shape of the model, do vainly glory and give out, that it is a

Masterpiece of their own finding out. As for myself, I must acknowledge, and willingly ascribe unto each man his due, and therefore will not blush or be ashamed to attribute justly my Philosophical principles unto my Master Moses, who also received them, figured or framed out by the finger of God; neither can I rightly arrogate or assume to myself the primary fabric of this Instrument, although I have made use of it in my Natural History of the great World, and elsewhere (but in another form) to demonstrate the verity of my Philosophical argument; for I confess that I found it graphically specified and Geometrically delineated in a Manuscript of five hundred years antiquity at the least . . .

CHAPTER III

First, we must observe that this our experimental Instrument is composed of three parts, whereof two of them are more essential and proper to the nature of the Engine or Machine; namely, the Matras, or Bolts-head, and the small vessel of water, into the which the nose or orfice of the Matras, after it is prepared, ought to enter; and the other is more accidental, as being only ordained to sustain the Glass firmly, in his perpendicular position, and to adorn and set forth the Machine.

Touching the Matras or Bolts-head, it is a round or oval Glass, with a long and narrow neck, whose orifice, or mouth and nose, ought to be proportionable to the rest of the neck, and it must be prepared after a two-fold manner: first of all, the long neck of it being put perpendicularly into the small vessel, being full of water, so that it do[es] touch the bottom of the vessel, we ought to measure from the superficies, or top of the water, and begin our division into degrees, still ascending upwards, till we come to the very ball, be it round or oval. And whereas the common sort of this kind of Weather-Glass has his first degree beginning downward, marked with the sign of 1, and so ascends upward to the round ball, according to the natural Arithmetical progression, thus:
1   2   3   4   5   6   7   8   9   10   11   12   13   14   15.

I, for a better methods cause, do alter the order in numeration, and dividing of the neck or pipe of the Matras in the middle, between the head of it, and the superficies of the water, I mark the place of the division with the figure 1 and so count my degrees downward and upward unto 7, after this manner: 7 6 5 4 3 2 1 2 3 4 5 6 7, which I affect, for reasons that I will express to you hereafter. So that the matter will be ordered thus:

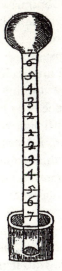

As for the small vessel of water, you see it here also described, with that proportion of the pipe of the Matras that descends into the bottom of it; which is so far from division as it enters into the water.

Now you have thus divided the neck of your Matras into parts, you must prepare and order it after this manner to make it magnetical and attractive by cold, and expulsive or dilative by heat; you must therefore take the orfice of the Bolts-head in your hand, and hold forth the head thereof, or the round which is above it against the fire till it be very hot; for the heat of the fire will rarify and dilate the Air in the glass, and cause by that means a good portion of it to fly out of the glass orfice, and so it will remain in that estate so long as the glass is in the degree of heat, at which time, if you suddenly put the nose of the pipe into the water, you shall perceive that as the Bolts-head does keel or was cold, so also will the water by little and little mount upwards into the neck of the glass. And we must note the hotter the glass is made, and the colder the external air will be found as that present, the higher and by so many more degrees will the water ascend into the neck.

And the reason hereof is, because that as heat does rarify the air, so the greater the heat is, the more excellent will the degree of rarefaction be. Contrariwise, as cold does condense and thicken, so the greater the cold is, the stronger will the condensation be, and therefore, after that the included air is much rarified, by an intense external heat, it follows, that as the heat does by little and little fade, so the cold will by little and little prevail, and have dominion; and consequently, the included rarified air must needs by little & little be condensed; but because there wants sufficient matter or corpulency in the air for the cold to work on, being that a part of it was spent and vaporated by rarefaction, it follows that as the succeeding cold does condense & contract the air, so the air, by contraction of itself, must also attract and suck up into it so much water out of the vessel as there wants air to satisfy the contractive appetite of the cold; for the interposition of the water between the external air and the internal or included air will not permit the internal to suck or draw into it the external to supply her want, for the satisfaction of the external colds lust, and therefore the water is forced to ascend, in lieu of the external air. And this is the mystery of the Instrument, and the manner of his fabric, whereby it plainly appears that they have been deceived which have deemed that the heat was the occasion of the attraction of the water upwards, being that each man may be an eye-witness that it is heat that drives it downwards; and contrariwise, that cold is the cause of his mounting . . .

CHAPTER IV

I divide as well the property as the use of this Instrument into two kinds, whereof I call the one general, and I make the other more peculiar. As for the general property of it, by the one it contracts and condenses, namely, when the included air is animated by the external cold; and by the other, it dilates and rarifies, to wit, if the included spirit be excited by any external heat. And therefore, through his constrictive nature or action, which is made evident by the contraction of the air, we may easily discern the universal reason of the inspissation and condensation of things that were thin. And again, by his dilation, we may scan or decipher the cause of rarefaction of such things as were thick. For by the speculation we shall find that there is nothing in the whole Empire of Nature which can be rarified and made subtle except it be by the action of light or fire, whether it be visible or invisible; and the essential effect of that action is light. And on the contrary part, nothing can be condensed or inspissated where darkness has not dominion; forasmuch as darkness is the essential root of cold, which is the immoderate action in condensation.

The particular properties, with the uses thereof, are manifold: for first, the nature of it is to discover the temper of the external air, or catholic element, in heat or cold; for the higher that the water does climb in the neck or pipe of the Matras, it argues that the firmer & stronger is the dominion of cold in the air; so that by this means we may daily judge of the increase or decrease of cold in the air; and by consequence we may guess at the proportion of heat in the sublunary spirit of the world, by the descent of the water.

Certain Experiments worthy of observation, and approved by many of this City, touching this Experimental Glass

If the water in the pipe of the glass, which before was highly mounted, does fall on the sudden by some degrees, it will be an undoubted sign that rain will immediately ensue.

If the water in the space of one night does descend, it is also a sign that rain will come not long after.

If the South or East wind do blow, immediately after a North or Westerly wind, the water will fall by certain degrees; but if the North wind or cold Westerly wind do blow, after a Southern or Easterly wind, then will the water be forthwith exalted.

If the water does attain to the figure 1, it argues that the Air is in moderation between heat and cold, as when the Sun is in the vernal Equinoctial, or as the natural temper of the Spring uses to be.

But if the water mount higher, then it argues that the disposition of the Air is by so many degrees more of Northern or Boral nature, as the water is mounted towards the Bolts-head; for you must conceive that the degrees from 1 to the uppermost 7 are belonging to the winter *Hemisphere*, and therefore are the degrees which note the augmentation of cold. So that if the water do mount up unto 2 in the Northern or higher part, it is an argument that cold has dominion over heat in the external Air only by one degree. If it mount to the 3 of the same Hemisphere, it foretells a slight frost, but if it ascend to 4 or 5 it pretends a hard and solid frost; if it come to 6 and 7 it argues great ice; but if it mount yet higher, it shows that a hard ice is likely to surprize and cover the whole river of Thames.

On the other side, if the water descend from 1 unto 2 of the lower rank of degrees which imports the Summer or hot Hemisphere, then it argues heat has gotten dominion over cold by one degree. But if it descend to 3 or 4 it imports a greater distemper of the Air its heat: if it descends to 5 or 6 it demonstrates the air to be exceeding hot, but if the water be beaten down to the lower figure of 7 it shows that extreme and Sultry heat, causing Coruscations and lightnings, has dominion in the Air.

So that we may discern how great a reference or relation there is between the external air or universal sublunary Element, and the Air included in the Instrument . . . I will in better terms express the Consanguinity and Sympathetical relation between one and the other in this subsequent chapter.

CHAPTER V

. . . Yea verily, and I aver boldly that the whole World, or worldly round, is as well and completely stuffed or filled with spirit or air, as is this our artificial vessel, or experimental

Machine; which if it should not be, it would consequently follow that vacuity would be admitted into the nature of things, the which would be but an absurd thing in a Philosopher to credit. Wherefore we may boldly conclude, that the spirit is in the like quantity, weight and proportion in the concavity of this Instrument, considering his magnitude, as it is in the great or little world. But experience teaches us that the selfsame nature, be it hot or cold, which uses to reign and have dominion every quarter of the year, in the cosmical or worldly spirit, does produce the selfsame effects in rarefaction and condensation of the air included in our artificial vessel ... And this is the reason ... that by the observation of this Weather-Glass that temper of the air in the great world is so exactly discovered to us ... [See Figure 3, p.98.]

I would in this regard have each discreet Reader to understand that when he beholds this Instruments nature, he contemplates the action ... of a little world; and that is has, after the manner of the great world, his Northern and his Southern Hemisphere plainly to be discerned in it, the which two are divided exactly by an Æquinoctial line in effect, which cuts the Degree signed with the character 1. Also it has his two Tropics, with their Poles; only we take the Southern Pole and Hemisphere to be hot, in regard of us, because the breath which comes from it is from the Sun, which in our respect is Southernly disposed; and therefore we term that Pole, the Summer-Pole or Hemisphere, and the other, the Winter-Pole or Hemisphere. And we have demonstrated that the degree in the neck of the Glass 1 does correspond exactly to the place of the Æquator, because that if the Northern or Winter Tropic be imagined to be the Basis of one Triangle, whose Cone shall end in the centre of the Northern Tropic, then is must follow that where the intersection is made by these two imaginary Triangles, the Æquator must of necessity pass. As for example [in the following diagram].

And we term the place of the Æquinoctial, the Sphere of Equality, because when as the Sun is in Aries or Libra, which are the vernal and autumnal intersection of the Æquinoctial,

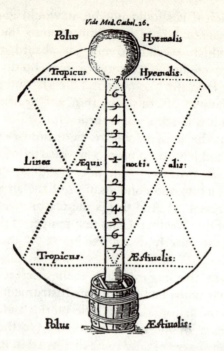

the days and nights are equal; so also the temper of each Hemisphere in heat and cold, is naturally observed to tend to a mediocrity or equality. Even such also will the temper of the micro-cosmical air, or catholic spiritual element be unto the Earth, when the water in the Glass is drawn up half way.

Book 2

CHAPTER IV

Of the false wisdom, spurious Philosophy and Philosopher; with their marks or characters

... for this reason therefore we find, in the one of the two kinds of wisdom, the fruits of power, virtue, and miracles, such as the true and divine Philosophers did produce by the Omnipotent Cornerstone (I mean the true Wisdom) in times

past, and made them manifest to the world; whereas the other
can do nothing indeed but produce cavillings, dispute,
contentions and fallacies, the fruits whereof, in the conclusion,
is naught else but vanity. It is not I, but the Spirit of Truth
that assures you thus much. And yet now, even in this later
age of the world, in which Satan, the prince of this world
which is darkness, has the upper hand; this terrene wisdom
or vain Philosophy, which is daubed over with dark ignorance,
has the dominion or upper hand, and so by the means Christ,
which is the true Wisdom, is daily crucified among some
Christian Philosophers, and buried in darkness, through the
misty and ambiguous clouds of that cavilling, brabling,
heathenish Philosophy, which they so adore and follow, with
their Master *Aristotle*, as if he were another Jesus rained down
from Heaven to open to mankind the treasures of the true
wisdom . . .

I could heartily wish, that each Christian Peripatetic, who
spends his time in disputes and cavils, after the *Aristotelian*
manner . . . seriously . . . call to mind that in the Church of
God, and habitations or kingdoms of the true *Sophia*, or, if
they please, *Philosophia*, there is no such custom as the Apostle
teaches us, for this mixtion of multiform human wisdom with
the wisdom of God has been the occasion so many dissentions
and discords, as have sprung up among the Philosophers of
this world, whereupon every kind of this false Philosophy has,
by stiff cavillations and disputations, maintained her Sect.
There also has been the occasion or errors in the Church of
God, as well among the Christians as Turks and Jews, for
amongst us Christians it has been the root of many Schisms
and Heresies, which have risen up in the re-search of one only
true God, which is the eternal Unity . . .

The Apostle does notable decypher or paint forth this kind
of Ænigmatical Philosophers of our Age [in margin: 2 Tim.
31] . . . *In the last days, shall come perilous times: for men shall be
lovers of their own selves, coveteous, boasters, high-minded, and proud,
&c. always learning, and never attaining unto the knowledge of the
truth. And as Jamnes and Mabres did resist Moses, so do these
withstand the truth, being men corrupt in mind, and reprobate*

concerning the faith. But these shall prevail no longer, for their foolishness shall be made manifest unto all men, as theirs also was. But thou hast fully known my doctrine, &c.

Now his doctrine was concerning the true Philosophy, whose foundation was *Jesus Christ, or the true wisdom and cornerstone, which sustains all, and is all in all, and fills all, and acts or operates all in all,* which is contrary to the tenor of the Ethnic doctrine, seeing that it makes an infinity of essential Agents in this world, as Dæmons, Stars, Elements, Meteors, Fire, Water, Cold, Heat, Man, Beast, Plant, Mineral, and such like, the which they will have as subalternate essential Agents to act and operate of themselves, not understanding that there is but one catholic and indivisible Agent in many mansions, which does operate by, and in, an infinity of organical vehicles, all in all, and over all. And this doctrine of theirs has so infected our Christian Philosophers, which are of their Sect, that they distinguish of Gods Being, saying that he is present *virtualiter*, and not *substantialiter*, or *essentialiter*. . . . who should say that Gods virtue can be without his essence or divided from his divinity, which is indivisible . . . so they dream of some accidents to be in God, which are distinguished from his essence . . . But I wonder, if that were true, how God can be said to fill all things, and operate all in all, if He be only the first efficient cause and not the general cause of all action in this world by His blessed Spirit, which He sent out into the world to do the will of Him that sent it, as well in Heaven as in Earth . . . What needs more words, when the Apostle in plain terms decides this controversy in the Text before mentioned? [in margin: 1 Cor. 8.5] *Though there be that are called Gods, as well in Heaven as on Earth, (as there are many gods, and many lords) yet unto us there is but one God, which is the Father, of Whom are all things, and we in Him, and one Lord Jesus Christ, by Whom are all things, and we by Him. But every man hath not that knowledge* . . .

And therefore seeing that the Aristotelian learning is contradictory in diverse main points to the positions of the essential wisdom or philosophy, a true Christian ought not in the main points to believe it, seeing that (as St *James* has it) the grounds and tenor of it are opposite to the Holy Bible, which is the only cabinet of truth . . .

CHAPTER V

. . . I beseech thee (gentle Reader) give me liberty, first to apologize a little for myself. Be thou therefore pleased in the first place to understand from my just and upright spirit, that I acknowledge and confess this Prince of the Peripatetic Philosophy to be a personage of a profound speculation; and that he had as deep an insight into the light of nature as any of the common rank of Philosophers in his time; yea verily, he had so sharp an ingeny [ingenuity], and so subtle and refined a spirit, that he not only allured by his worldly craft and human invention the Gentilish Greeks (whereupon he was termed by them, Cacodæmon, or a deceitful spirit, or seducing from the truth), but also Christians themselves of every sect, even to this present, insomuch that they are so wedded to his worldly wisdom that they admire each new proficient in Philosophy, of their Universities to maintain his actions, and not to decline from his doctrine. As for myself, though I may be ranked in that number, yet now I have collected my spirits, and have by Gods grace attained to that light of Holy Scriptures whereby I am made able to distinguish and discern their essential colours for this præstigious one of Pagans . . .

Aristotle was before the incarnated Word, so also is it evident that he knew little of the Mosaic learning, which consists upon the Creation effected by the spageric of the divine Word, when he would have the world to be eternal. I confess that his Master *Plato* was more essentially grounded on the true wisdom; but *Aristotle* being puffed up with self-conceit, would, in derrogation from the Stoical doctrine of his Master, arrogate all wisdom to himself by framing out or fashioning a new worldly wisdom or Philosophy which was afterward termed *Peripatetical*; and so by his vainglory, he added to some truths many of his own inventions, making as it were a Gallimofry of good and bad, of true and false, of wisdom and folly together, which is far from the nature of the perfect Christian wisdom which must needs be therefore wholly truth itself, because it is described by the Spirit of God, in the which there is nothing but truth . . .

. . . why should we be at strife, disputes, and brables, about difficulties or ambiguities or questions? or, why should our riper senses be battered in pieces by foolish distinctions in which are the inventions of human or diabolical wisdom, on set purpose to immerse and drown us deeper in the abyss of multitude, or profundity of ignorance, when the only endeavour of truth is to conduct us through those clouds of errors . . . to the fountain of Unity and Concord, which is the eternal Wisdom, the spiritual Christ Jesus? Thus we see how contrary the Greekish wisdom is to that of the Apostles; forasmuch as the Greekish Philosophy is contentious, litigious, full of disputes, brables and emulations; for which reason it is pronounced by St *James* to be *terrene, animal, diabolical and not from above* [in margin: James 3]; whereas contrariwise that Philosophy which is grounded on the true wisdom, which is from above, *must be first pure, then peaceable, gentle, easy to be entreated, full of mercy and good fruits, without judging, envy or hypocrisy* . . .

Book 3

CHAPTER I

It is an evident Argument that the Ethnic Philosophers were not well settled upon the grounds of their Philosophy, but did waver in their imaginations touching the true principles of Nature, being that among each Sect of them there was maintained and upheld a variety, yea and sometimes a plain contrariety of opinions concerning them . . .

Now the main error of these Philosophers in their judgements concerning the principles was that they did not mark or consider that the divine puissance or sacred word was more ancient, and of a greater Antiquity, than were any of their foresaid principles; the which, if by a riper contemplation they had understood they would have confessed, being instructed and directed by reasons produced from the eternal unity . . . that the divine light or sacred emanation (which Scriptures entitle by the name of the Holy Spirit of wisdom) was the actual beginning of all things . . .

And yet it is most apparent that some of the Greekish and Ægyptian Philosophers, namely, *Plato, Pythagoras, Socrates, Hermes,* &c. did so instruct their understandings, partly by the observation of their predecessors doctrine, and partly through the experience, which in their long travails and peregrinations they had gathered, among the Æthiopians, Ægyptians, Hebrews, Armenians, Arabians, Babylonians and Indians (for over all or most of these Countries did *Plato, Pythagoras, Hippocrates* and others of them travel for the augmentation and increase of their knowledge, as Historiographers, that are worthy of credit, have related) that without doubt they did discern, though afar off, and as it were in a cloud, the true light in the humid Nature. And among the rest it is reported, as also it appears by his works, that *Plato* had the knowledge of the Word, and had read the Books of Moses; and for that reason he was called *Divinus Plato,* the divine Plato. In like manner, the excellent Philosopher *Hermes,* otherwise termed *Mercurius Trismegistus,* expresses plainly that he was not only acquainted with *Moses* his Books, but also was made partaker of his mystical and secret practice, as by his Sermons, which he calls *Pymander,* a man may plainly discern, where he mentions the three Persons in Trinity, and shows the manner of the worlds creation, with the elements thereof, by the Word. And therefore of all the other ancient Philosophers, I may justly ascribe divinity unto these two.

But in this I cannot much commend them, *viz,* in that they having had a view of *Moses* his labours, which were indited by the Spirit of God, did gather out and confess the truth of his doctrine touching the principles of all things, and yet would not in open terms acknowledge their Master, but altered the names of them; but as *Plato* served his Master Moses, even so was he dealt with by his scholar *Aristotle,* who knowing that his Masters three Mosaicall Principles of things, masked under strange titles, were but truth, would nevertheless arrogate his doctrine to himself, and for that cause did alter the assumed names of *Platos* principles, gilding them over with new denominations and did afterward rear up upon them a spurious Philosophical structure, carved and framed out after his own inventions . . .

And yet as for *Plato* and *Hermes*, I must excuse them, being that they do both of them acknowledge in express terms with *Moses*, that the matter or substance whereof the Heavens and the Earth were made was a humid nature, and the internal form or act which did dispose of it into diversity of figures or forms was the divine Word, as you may find most plainly expressed in *Platos* works, and in the *Pymander* of *Hermes* or *Mercurius Trismegistus*. But amongst all the rest, Aristotle has soared highest upon the wings of his own conceited imagination, and built the structure of his worldly wisdom upon the typical form of the Mosaicall grounds, thinking thereby to assume and purchase unto himself, in the regard of this world, the name of an absolutely wise man, though in the conclusion he appears for otherwise in the eyes of God, for as much as he does assign particular essential actions, which appertain really unto God, unto the creatures, with more obstinacy than the rest, affirming that they operate essentially of and by themselves, when in Verity it is only God that *operates all and in all*, and that immediately . . .

And this is the reason that they give not unto God (the only Creator) the glory of every action in this world (as they ought to do), but rather to a created Nature, and unto Angels, and Stars, and Elements, and compounded creatures, which were made, and are still sustained and maintained by the all-creating Spirit or word of the Almighty. And this is the original occasion of the multiplicity of Idolatry, which has and does hitherto reign in this world, namely, or the worshipping of the Sun, Moon and stars, of sacrifices offered to Idols or false Gods, and deceiving Devils, of the Veneration of *Isis* and *Osiris*, of the adoration of Saturn, Jupiter, Mars, Venus and Mercury, of the immolations or offerings to *Cælum, Vesta,*

Ceres, Proserpina, Vulcan, Pluto and *Neptune* . . .

CHAPTER V

. . . it is necessary for us to know the property and sense of *Plenitude and Vacuity*, according to the true Wisdom or Christian Philosophy. And first, I will speak a word or two of that Vacuity or Emptiness which is so detestable and odious in the works which the Creator hath made. As well the Fathers of the Philosophers . . . have termed it by the name of *Nihil* or *Nothing* . . . *Vacuity, Inanity, Nothing and Darkness* are one and the same thing: to wit, Vanity, Inanity or Voidness, because that all fullness and plenitude is from God in his actual property. But God did not as yet shine forth unto the world, and therefore as the first deformed matter of the world was void and destitute of all inacting grace and formal goodness, as was said to be *Vain, Empty, and Darkness* . . . Wherefore the Earth that was before the revelation of Gods Spirit inane and void, is now become full of divine Light and multiplying Grace. Whereupon it was no more void and empty, that is to say, destitute of essential being, but became fertile and fruitful, being now replenished with divine fire, and the incorruptible Spirit of God . . .

Whereby we may perceive that all plenitude is from the divine Act, as contrariwise Vacuity is, when that formal life is absent from the waters, and this is the reason that *Vacuum* or *Inane* is held so horrible a thing in Nature, forasmuch as the utter absence of the eternal emanation is intolerable to the creature, because that every thing desires fervently to be informed, and that by a natural appetite and affection, and therefore it is abominable to each natural thing to be utterly deprived of being. For this reason it follows that unless God had filled all things in this world with his Spirit, *Vacuity* and *empty deformity* would have possessed the world . . . Therefore it follows that there is nothing in this world which is Inane, in vain, or void and empty . . .

CHAPTER VI

... we must imagine that when there was no formal Light to inact the deformed Abyss or Chaos, and consequently no Heat to act and make a division of Light from Darkness, then all the dark Chaos was inclining to drouth and wet, congealed with cold: for Cold has an infinite power and dominion where Heat is absent, as we see about the Northern Pole, all the waters are frozen into a dry and cold clod or heap, by reason of the absence of the Suns Heat. In like manner where the Sun, or burning Easterly wind, does send forth directly their spiracles of Heat, there Cold is banished, and the Earth is turned into a dry, thirsty and spongy mass. Even so and after the like pattern ... the Chaos was a confused, cold and dry heap, until the Light did appear, and began to operate. Also the Light of itself, as it was considered without any action upon the passive mass, was of a fiery condition, that is to say, Heat and drouth did only appear in it: [it] follows therefore, that as cold in effect is nothing else but the act of the divine puissance made potent and evident by the absence of Heat, so also is drouth or dryness nothing else but an apparent passive nature, evermore accompanying the two foresaid active powers in their absolute intention, where moisture is totally absent. Now moisture is as it were a mongrel, begotten between the two opposite actives, which is proved thus. The North wind by his dominion turns the Air into Snow, Hail, Ice or Frost, that is, into a cold and dry Substance. But when the Southerly or Easterly wind does begin to have dominion, then their blasts do penetrate, by little and little, the said dry masses or substances, and undoes them, and converts them into a moist or humid nature ... So that moisture is nothing else but a mixtion of Heat and Cold, in one solid Substance, which is fluxible *in potentia*, and then by little and little does the cold and drouth depart, and become faint, according as the power of the wind is more or less vigorous in Heat. For thus much we must observe, that as Cold does make immobile and fix[ed]; so it is an evident sign that the Southern or Eastern Heat has taken possession of the Mass when by their action it begins to moisten, to revive and move again . . . In the very same manner also, the uncreated and all-creating Spirit of Light, moving upon the face of the dark abyss, did operate in it, and made the congealed Mass to relent, and then it was called by *Moses*, waters; and by *Hermes*, *humida natura*, or the *humid nature*. And so they continue to this day, being altered from one nature to another, according to the will of God, which he effects by his Ministers, the Angelical winds, causing the Spiritual waters to change and wheel about from one nature into another . . .

An ocular Demonstration, confirming the Divine properties or Virtues above-mentioned

Before we proceed to our ocular demonstration, which shall be made in and by our Experimental Instrument, we must consider in the first place that the Catholic Air or Sublunary Heaven is the subtiler and more spiritual portion of those waters which are under the Firmament, of which division Moses makes mention, and therefore every particular thereof does correspond to the whole, and consequently the air included in the glass of the Instrument is of the same continuity as the excluded whole fares, so also does the included part. Again, as the Spirit which walked upon the waters did animate, vivify, inform and dilate them giving them motion, so also by his absence, or by hiding its act or contracting its emanating beams into itself, the waters are also contracted, condensed and darkened . . . But as the waters do by their existence fill the vaulted cavity of the world, so also does the all-informing Spirit fill every corner of them, insomuch as being it operates all in all, but in a diversity of property . . . [See Figure 3, p.98.]

Seeing then that it is water that is the catholic passive . . . and that the eternal creating and all-enacting Spirit of the Lord is the universal actor which moves all in all in the waters . . . by an infinity of Organs, as Angels, Sun, Moon, Stars, Winds, Fire, &c . . . it must needs follow that He is the agent, as well in the contraction and dilation generally, without the Glass, as particularly within the Glass. Wherefore as the Sun, the hot winds, the fire, or natural heat of mans body have their dilative property from His emanating and enacting virtue, and do alter by it the cold air, the winds and water from his privative disposition.

So it follows that as well the dilation of the air in the Glass as contraction is the immediate act of this Spirits positive or privative property; for when this Spirit blows from the North or West, the air is contracted more or less into a narrower room within the Glass, and that is proved thus, namely, because the water is drawn up higher into the neck of the Glass, lest a corporal vacuity should be admitted in nature. And again, it is most apparent, that the air in the Glass [is] by so much the more contracted by how much the Northern cold has dominion in the outward air, because it is gathered into a more straight place or passage than it was before the water was elevated up. On the other side, if the hot winds, or Summer Sun, do inflame the external air, then the included air will also dilate itself, and in its dilation require a larger space. That the air is so dilated by heat it is plainly demonstrated, in that the water is struck down by so many degrees lower than it was. Again if one put his hand on the top of the ball of the Glass, the water will sink, for the air will forthwith be dilated.

Now that the spirit of life, which gives this heat to Man or Beast, is from this eternal Spirit, which (as the Apostle says) *does vivify all things*, Scriptures do in many places, above and hereafter mentioned, confirm. Again, the Prophet calls this one *Spirit from the four winds, to breathe into the dead carcass that they might live again* [in margin: Ezek. 379]. Thus you see it evidently confirmed by an ocular demonstration that cold does contract, inspissate, and make gross the included air, which is argued by the drawing up of the water and straightning the air. And again, that heat does dilate and dissipate, by the enlargement of the air in hot weather, or by laying of the hand on the Bolts-head, which is made evident by the beating down of the waters.

Note (I beseech you), ye that will not be over-partial on the Peripatetics behalfs, the two notable errors of the

Aristotelians: whereof the first is manifested, in that they hold for a Maxim, that *heat does congregate and gather together things of one nature.* Now you see it here ocularly demonstrated, that heat does operate the contrary ... But it is cold that does congregate, compact, and gather homogeneal things together as well as heterogeneal ...

The other of their palpable errors is also described by this ocular experiment, for whereas they say that the Sun, stars and Fire, yea, and all heat whatsoever attracts and draws to it the vapours and humidity of the earth, water, &c. we find the contrary by this our experimental Glass; for in only laying the warm hand upon the Glass, the air dilates itself immediately, as is so far from being attracted by the heat, that contrariwise it flies away from the hand ...

The Fourth Book

CHAPTER I

I have made it (as I imagine) most plain and evident to your understanding, as well by the testimony of the antique Philosophy, and infallible wisdom of Holy Writ, as by ocular demonstration, that the common or universal matter and subject of all things, was the Waters, which were inacted and created by the bright Spirit of the Lord, before all things; which being so, and for that all Condensation and Rarefaction do regard a subject or material substance, wherein it should be performed or effected, it follows therefore, that the only matter which does endure or suffer the act of either of them, must be water, or a humid and moist nature, being it is the only substantial stuff which fills all the vast cavity of the world, and consequently the material existence, of which both Heaven and Earth, and all things therein, are framed, and were originally shaped out. This is therefore that main subject of Condensation and Rarefaction, by the means whereof all things in this world are made to differ from one another and are disposed and ordered by God according to weight, number and measure, in their proper ranks and places; that is, they are placed in a higher or lower region according to that dignity which the catholic or eternal Actor has bestowed upon them in their creation. The common matter therefore of incrassation and subtiliation being thus made manifest, we are to find out the universal actor or operator in this work. And I did signify to you before that it was the sacred emanation of that divine splendour which did operate from all beginnings and does at this present operate . . . in the Heavens above and the Earth beneath . . .

And therefore he must needs act all in all, according to his will and pleasure. Again, when he is pleased to withdraw his actual beam within himself, he seems to rest, and the act of the creature does cease, and then death ensues . . . even as we see in the Sun . . . that the beam is continuate to the body of the Sun as the emanation from the divine fountain is one individual essence with the fountain of eternal light . . . even as we find that the Holy Spirit which is emitted from Father and Son is one in essence with them both . . . I conclude therefore, that Rarefaction is nothing else but the dilating of water by heat, which was first contracted by cold; and Condensation, on the other side, is a contraction or sucking and drawing together of those watery parts which were dilated or dispersed by heat, so that the subject of both these actions is water, and the fountain, as well of the privative as positive agent, is the Nolunty or Volunty, that is to say, the negative or affirmative will of that one eternal Unity who is that all-informing and operating Spirit which acts and accomplishes his pleasure in all and over all by his diverse organs, as well angelical as celestial and elementary, which, according to their diversity in disposition, are moved by this internal act to operate in this world the will of the Creator, both positively and privatively . . .

CHAPTER II

The Eternal spirit of wisdom, who is the initial principle of all things, and in whom and by whom . . . the *Angels, Thrones, potestates and dominations were Created* [in margin: Colos. 116], does operate by his Angelical Organs of a contrary fortitude in the Catholic Element of the lower waters; both the effect of Condensation and that of Rarefaction. And to verify this out of the Holy Bibles testimony, we read first that this one spirit is the arch-Lord and Prince of the 4 winds, or else the Prophet by the commandment of God would not have said; *Come O spirit from the 4 winds and make these dead Carcasses to live again* [in margin: Ezek. 37.9]. Now that this one spirit works in and by spiritual and Angelical Organs in the execution which is effected by the property of the 4 winds, it is proved out of the *Apocalypse*, where we find these words: *I say 4 Angels stand on the 4 Corners of the Earth holding the 4 winds of the Earth*, that they should not blow on the Earth neither on the Sea, &c. By which it is evident that these Angelical Presidents over the 4 winds were emitted or restrained according to the will of that eternal spirit, which guides them when and where he list.

But we find by daily observation that the essential virtue in the Northern spirit is cold, and therefore contractive or attractive from the circumference to the centre, and by consequence a causer of congealation and condensation. By this kind of Angelical virtue, the divine spirit does work his privative effects, and acts of Inspissation and fixation in the sublunary Element, causing terrestrial and earthly effects. But Contrariwise, we find that by and in the Southern and Easterly Angelical spirits, the all-Creating and operating power does cause Rarefaction and Subtiliation in the said inferior waters through their essential action, which is heat. And for this cause the said hot winds do undo by Rarefaction all that which the cold Northern winds did effect by Congealation.

Hence therefore it comes, that of this potent Angel (who is said by the wise Cabalists to be the President and Governor of the Celestial Sun, which some term *Michael quasi Quis sicut Deus*, of whom we will speak hereafter, also touching his Angelical organs in the 4 Corners of the Earth, by which he uses to do his will) it is said, *The Son of Man will send out his Angels from the four winds of Heaven* [in margin: Mar. 24]; and moreover, we shall find in the place before mentioned, that this imperial Angel did rise from the Easterly angle of the world, and did command the foresaid 4 Angels which had dominion over the 4 winds, as vassals to his will. For the Text has it: *And I saw another Angel come up from the East, and cried with a loud voice unto the 4 Angels unto whom power was given, &c. Hurt not the Earth, neither the seas, nor the trees till I have sealed, &c.* [in margin: Apoc. 72]. Whereby it is apparent that the 4 Governors of the winds are subject to that great Angel who is the powerful and potent Intelligence which moves and rules not only the Sun, but sustains all things by virtue of this word ... the Sun by virtue of this omnipotent Spirit, whose immediate angelical organ or Instrument the Angel *Michael* is, does govern the airy spirit, both of the lower and upper world ...

Whereby it is evident that the eternal Breath is that which animates the Angels; the Angels give life and vigour, first, to the stars, and then to the winds; the winds first inform the elements, or rather alter the catholic sublunary element into diverse natures, which are termed Elements; and then by the mixtion of diverse windy forms in that one element, they do produce meteorological compositions of diverse natures, according to the diversity of the windy forms which alter it ...[1] [See Figure 9, p.176.]

CHAPTER V

... whereby it appears that at the blast of [the] easterly winds, the air or catholic element becomes burning and fiery, so that it heats and enflames the bodies of the creatures. On the other side, the occidental winds are found by experience to be opposite in nature and condition unto these; for they convert the hot air or general element into cold and natural visible water, being that they are the procurers of cold rains, so that we may see by this, that the formal act in each Angel of the four corners of the Earth ... has an essential virtue imparted to it from God, according to his volunty, at the instant of the Angels information; and therefore in that very property, does the angelical creature act in the common element, or lower waters, in the which the creating Spirit properly was, when he made it, so that the common element is daily informed anew and altered by essential act of the angelical wind which blows with dominion; insomuch as if the easterly angelical wind informs it, then it becomes a fiery element, for it heats and dries by super-excellency: if a southerly, it is changed into that elements nature, which is called air; and if a westerly spirit has dominion, it is converted into the temper and proportion of water; lastly, the Northern blast transmutes it to the consistence and disposition of earth . . .

The Demonstration

We find that from the lower region of the Calendar- or Weather-Glass to the summit or top of the head there is nothing but a portion of the common invisible element, namely, the Air within it. But we shall find even in this little model of air, strange mutations or alterations effected by virtue of the four winds which blow in the open element; for when the hot Easterly wind does blow, it dilates and extends itself all along the neck of the Glass, and beats down the water to the lowest degree by reason of its extension, so that it approaches the nature of fire; for fire is said to be nothing else but air extremely dilated . . . but if the South wind blows, then it will not be so extremely dilated, but will endue the mean nature of air, and therefore it will draw up the water by certain degrees. But if it happen that the Westerly wind have the sole dominion in the sublunary element, then will the air in the Glass grow thicker, and for that reason it draws or attracts the water higher . . .

Lastly, if the cold Northern winds do govern or dispose of this universal sublunary element, then will the include[d] air be contracted . . . into a very straight room . . .

Thus you see that in verity there is radically but one catholic sublunary Element, though by the angelical spirit that blows from the four corners of the Heavens, it is four-foldly informed and altered . . .

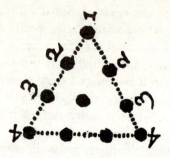

To conclude, I will demonstrate the mystery of the worlds Creation, by way of an Arithmetical progression, after [the following] manner.

Here we have the progression in the worlds creation, where 1 signifies the Unity which was before all things, which while it was in itself, and did not shine forth, contained its inacting property within its potential *Nolunty* or Darkness, and therefore was esteemed as nothing, in regard of mans weak capacity, although that in itself, it is all in all. Then the rank 2 shows the actual emanation of the un-created Light, out of the potential Unity, at the presence whereof the humid nature did appear out of the dark abyss 10 in the shape of waters so that the light and waters as active and passive are ranked next after Unity thus,

0.	Light
0.	Water
2.	

Then by the fiery Spirit of eternal love and union, those two opposite natures are joined together into the nature of Heaven by a spiritual union or composition: the which is termed by the Platonics, the soul of the world, which the Philosophers have styled by the name of *Quinta essentia*, the quintessence. For we must understand that as the 4 Elements were made after the Heavens, so also are the Heavens said to animate the Elements, no otherwise than the soul does the body. So that this degree of Progression in the Creation stands thus,

0.   Light
0.   Air
0.   Water
3.

The last scale in the Progression of the worlds Creation is thus

0.   Fire
0.   Air
0.   Water
0.   Earth

which imports the four-fold alteration of the catholic Element by the four Winds, which was and is effected by the Word the third day of the Creation, and this was nothing else but the general sublunary or lower waters.

... this I say, is the manner of the worlds Fabric, as also of the rotation of one Elementary nature into another, cased partly by the absence or presence of the Spirit, riding in his chariot or Tabernacle, which is the Sun. Hence therefore proceeds that alteration by Condensation or Rarefaction which is observed annually in the world, but especially by the 4 windy Organs, or Angelical Instruments of this aerial region, and is effected assidually by changes all the year long, as is justified by the Weather-Glass observation. And we must with diligence observe that these members of the worlds Fabric do endure with incorruptibility always, and shall never alter until the last PEREAT.

But the creatures which are compounded of this general Element, and are diversely altered or informed, shall pass or begin their generation from the simple Elementary estate, which is in 4, *quasi à Termino à quo*, unto the complement of generation, or complete composition, namely 8 *tanquam ad Terminum adquem*. And again, the corruption or resolution of that generated compound shall have its *Terminum à quo*, or beginning from the degree of complete composition 8 and his *Terminus ad quem*; that is, the complement or resolution shall be in the common or catholic Element, which is air, four-foldly altered in his simple nature ...

CHAPTER VI

. . . now therefore that the progression or ranks of the worlds Creation and his simple Members is showed you, which consists of 4 ranks, in which estate the world must (as I said) endure till it be refined with fire, I will make my progression from these simple members of the worlds existence, and proceed unto the order of degrees in composition, which have their beginning and foundation from them, namely from the first degree or rank in Generation or Composition, unto the Complement thereof. And as between the beginning of mixtion, and the perfection thereof, there must intercede a medium or mean, which must be . . . an imperfect mixtion: so betwixt a simple Element and a perfect body, an imperfect composition must needs intercede, namely such a one as is that of a Meteorologic nature.

As for example, in the generation of man, the beginning, which is founded on the catholic watery element, and takes his place in the first rank, is Sperm; for it is a watery or fluid Substance, but little altered, and as in the water the whole Fabric of the world, and seeds of all things was complicitly contained, and yet nothing did appear externally but water. So in the seed or Sperm, though nothing do appear in the first degree explicitly but Sperm, yet the whole man, namely the bones, flesh, blood, sinews and such like, are complicitly contained, and will by degrees appear out explicitly, namely in the sixth rank, for it will be altered from humour to solidity, with a certain distinction of the three principal Members [then] in the seventh, to an Embrional shape, and in the eighth which will make up a cube unto the material root 2 or the Square 4 (which is characterized in the catholic Element by the impressions of the four winds) into a perfect creature.

In like manner, in the great world we see that the simple Element, namely the general air, appears externally plain simplicity, and an invisible Nothing explicitly, and yet it contains complicitly, a cloud, water, or rain, Fire or Lightning, and a ponderous stone, with Salt and such life, which by degrees do explicitly appear through the Virtue of the four

Winds. So that a vapour possesses the first rank, the cloud the sixth, the Lightning and cloud the seventh, and the earthly stone argues an exact rotation of all the 4 ventous forms into one mixtion, which represents the eighths place in Composition or Generation.

But when the man comes to Corruption, then his parts proceed in resolution backwards, namely from 8 to 7 from 7 to 6 and from 6 to 5 until it return unto the point of the simple Spermatic Element from whence it began, and there it begins a new Generation in another form. For the all-acting nature

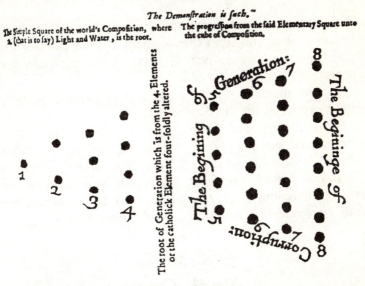

The Demonstration is such.

The Simple Square of the world's Composition, where 1 (that is to say) Light and Water, is the root.

The progression from the said Elementary Square unto the cube of Composition.

is never idle. So also the Stone is resolved into water, and water into a vaporous cloud, and the cloud into air from whence it came, which is the simple catholic Element which admits no farther or profounder resolution by corruption.

Now the only Operator in both these works is the Spirit of God: for in Generation it shines forth of the catholic Elements centre or Sperms internal [centre] unto perfection and perseveres in his action, till a perfect man be produced . . . Contrariwise, when the Spirit of God withdraws His beams from the circumference of Generation and Composition to the centre of simplicity, He leaves to visit the Spirit of the creature,

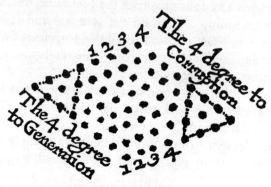

and so it must fade, and decayingly return to the principle from whence it came; and from thence again, if the same spirit is pleased to shine forth, a new Generation begins, where the Corruption or Resolution ended.

Or, the manner of generation and corruption is more plainly expressed [in the diagram above].

Where the four elements remain in their simplicity, as they were created complicitly in one watery nature, or rather catholic element Air, which is the root from when generation arises to the period of perfect composition, by four degrees or steps of alteration, namely, from the 4 to the 8 and whither tends retrogradely corruption, namely, from 8 to 4.

SECOND SECTION
Book One

CHAPTER IV

. . . but that I may the better express to you by a Philosophical Demonstration the Angelical nature of this supreme intelligence, called *Metattron*, and consequently the soul of the world; I would have you in the first place to know the true manner how the Philosophers do demonstrate the harmony of the world, and his spirit. The Platonics *Heptachord*, the which he did invent and adapt for the demonstrating of the soul of the world, did consist of seven strings, or proportions, partly even, and partly odd, namely 1, 2, 3, 4, 8, 9, 27. The which proportions, although *Porphyry* and *Proclus* have drawn forth in one line; nevertheless it appears to me that *Adrastus* and *Calcidius* have more fitly expressed and adapted it to the fides of a *Pyramis*, or Triangle, in this manner:

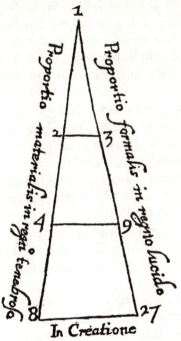

In the summity or top whereof, namely where both lines meet in one point, the figure of 1 is expressed with Unity, because it participates both on the side with the material existence and on the other it has a commerce with the formal emanation . . . The six other figures which do correspond to one another, from the material side to the formal, namely the even numbers that regard directly the odd, namely the material number 2 regards the formal 3; the 4 the 9; and the 8 the 27 do most lively express the general kinds of all creatures, with their harmony . . . For first, after the example of the *Archetype* from 1 issued 2 which is termed by the Pythagoreans the confusion of Unities, and therefore it is the radical or incipient imperfect number . . . and for this reason also, the watery matter that issued out of it is of itself imperfect, no otherwise than the number of 2 is esteemed in regard that of 3 because all perfection consists of 3 terms, namely a beginning, middle and an end. So that if you take one of the Unities of 2 for the beginning, and another for the middle, then there wants an end ... And this formal Unison is Symbolically expressed by the figure 3 which for this cause is termed the first number of perfection ... and according as in the Archetypical world, designed by 1, 2, 3 in form of the sacred Trinity there appears to be three Ideal dimensions in one divine nature or eternal essence, which present in their manner of progression a Root, a Square, and a Cube, which import a supreme soul, a spirit and body of the world Ideally painted out in the Archetype ...

... wherefore we must imagine that the multiplication of matter in the world is nothing but variety of penetration of formal light into the watery abyss or material multitude, for the thicker matter is, the nearer she appears to her root 2 or the dark Chaos. Again, the thinner it is made by the opulency of the formal emanations bright presence, the nearer it appears to the root of formal Unity. This mystery (I say) being rightly understood, all science, even in the abstrusest Philosophy, may easily be deciphered.

... by the multiplying of 3 in itself, we have the created formal square, which imports the nine Orders of the Angels, which vary according to the multiplicity of properties, by which the effluxion or emanation, that is sent out from this infinite fountain of light does operate diversely in the world, according to the will of the Father of lights, which sent it forth. And the formal square 9, being joined with the material square 4, does animate the Heavens which ... are made corporal or cubical, so that all bodies are made of those thin substances, by multiplying of each square in his root, whereupon the cubical form will be 27, and that of matter 8, which argue every elementated or compound matter; and as the more the matter is multiplied in itself, the darker and thicker will it be; so the more that light is multiplied in it, the lighter, thinner and more spiritual will the creature be. Insomuch that the formal root, and square and cube import the essence and souls of creatures, as well super-celestial as celestial and elemental, which are more or less dignified with form; for according to these multiplications in form, the more will the creatures be exalted in excellency.

You see now how far, and by what proportions, as well spiritual as material, the Platonic harmony of the world extends itself, and may observe, that where this harmonious proportion between form and matter is not, there must needs be as well spiritual as corporal dissention, or discord, and consequently antipathy. We see also, that the root of life is fixed in the angelical composition, which is of simple, light, & pure spiritual matter, so that the eternal sapience, or essential soul, is the act of the Angels; the ærial angelical spirit is the act or essential life of the stars, or heavenly influence; and the starry influence is the soul and life of the winds, the spiritual emanation from the winds do four-foldly inform the catholic sublunary element, or lower waters; the element does animate the meteorological impressions, and of these are the compound creatures compacted, which draw from the divine fountain of them all, being one spirit in essence, but multiform in regard of the variety of organs, by which it works variously in the world.

So that it appears that God animated immediately the Empyreal Heaven, or the intellectual spirit, which is the seat of Angels, and this we compare to the root; the Empyreal Heaven animates the stars, or ætherial region, which we refer to the square; and the starry Heaven is, as it were, the male, or multiplier and vivifier of the elementary region, and his creatures, which we compare to the cube.

. . . and therefore the learned have founded on this subject their formal and mystical *Arithmetic* and *Geometry*, which are not exercised about common and vulgar subjects, but wholly employed about the profound meditations of the true *Cabala*, natural *Magic*, and essential *Alchemy*, which because the ignorant vulgar people do rashly condemn under those titles, are otherwise termed by the mystical with the name of the science *Elementary*, *Celestial*, and *Supermundane* . . .

I could also demonstrate that the world and his soul, or

life, was shaped after the image of the Archetype in this manner: from 1, which was all light, in whom is no darkness, did 2 issue, which was darkness, or the dark Chaos, so called, because unity did not as yet shine forth to inform them. Betwixt these two extremes is 3 interposed, as a peaceable or charitable unity, between these two principles, [which] unites the divine formal fire with the humid material nature, or spirit of darkness, making a union of two opposite natures, so that both natures do remain in one sympathetical concordance . . . I express them [in the following diagram].

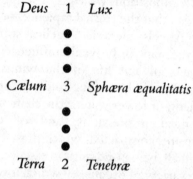

This excellent harmony of the soul of the world is elsewhere most fitly expressed by me by two equal Pyramisical shapes, whereof the one is formal, the other material; the basis of the formal is the immediate act of God, or the infinite and only bright Unity, and it imports the emanation of the creating and informing Spirit, so that his Cone does penetrate to the very centre of the dark Earth or abyss. And contrariwise, the basis of the material or watery Pyramis is in the Earth or centre of darkness, and the Cone ascends to the centre of the basis of the formal Pyramis, in [the following] manner.

Now where the intersection is made between the two principal concurrences, I have framed a piece of a circle, which circle, because the portions of the formal Pyramis, and material, are there equal, we therefore with the Platonists do call *Sphæram æqualitatis*, or *the Sphere of equality*; or as they in another respect term it, the orb or sphere of the soul of the world, which is just in the midst of the starry Heaven, called

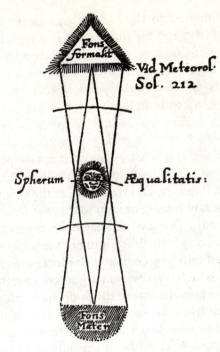

for that reason Æther, *quasi igneus aer*, a fiery air, as who should say, an equal portion of the spirit of the eaters, and of the formal fire descending from God or Unity (as I have expressed before). And therefore as we placed in the precedent demonstration the 3 in the centre or middle of the line, between the divine illuminating unity 1, and the dark Chaos, or deformed duality 2; so in the world was the body of the visible Sun of this typical world placed, in which the invisible and uncreated Sun of the Archetypal world did put his Tabernacle: and for this cause it is rightly termed of the Platonists, the sphere of the soul ... [See Figure 4, p.99.]

Now that this most excellent and perfect concord of life does remain principally in the midst of the line, drawn from Unity, or the fountain of form, to the Earth or duality, which is the fountain of matter, I prove it thus by the accords of Music: the perfectest consonance in Music is *Diapason*, and of all the other symphonical accords, it approaches nearest to the sacred Unity in Divinty, for it is half unison, and therefore it is rightly compared to the blessed emanation of life which came from Unity; and for that reason it is rightly termed the Idea or image of a unison. And as the unison like one essence in Divinity does comprehend three Persons, importing the three accords in the Archetypical harmony, so also the *Diapason* comprehends in it the two other accords besides itself, namely, *Diapente*, and *Diatessaron*. Now that this most perfect consonant of *Diapason* is planted in the midst of a unison, and does, as it were, beget itself in two perfect chords, or *Diapasons,* which be contained within itself, I prove it thus: take a Lute or Bandora, and strike any of the strings open, and then stop that same string exactly in the middle of it, and each half will sound a *Diapason* to that open unison; so that the unison which is made by the striking of the whole string open, will be divided into two *Diapasons* or most perfect consonants to the whole; which is an evident argument that the perfection of the *Diapason* is in the seat of the Sun, and consequently, that the formal and lively accord of all the world, is no otherwise in that central or middle place than the heart which is the seat of life, is in man. And yet by the beams of the Sun, this life of the world is made catholic, and fills all, no otherwise than the vital blood does universally expand itself in the Microcosm by the channels of the arteries . . . [See Figures 5, 6, 7, pp.171, 172, 173.]

The Conclusion

. . . it is not my intention to express my more bold and settled opinions touching the true Philosophy and his appendixes in a larger English phrase, considering the roughness and harshness of my pen, by reason of my defect, and the insufficiency in the polished nature thereof . . . I do imagine, and my mind gives me, that although I be never so curious in my inditing, or laborious in the phrase of mine expression, yet will my best endeavours appear faulty in the curious eyesight of some men, though perchance acceptable enough unto others. I esteem it sufficient therefore, that I dare be hardy and bold in the fundamental subject of the essential Philosophy, being that it has Truth itself to maintain and defend it, without any adornment made by the gilded tresses of superficial speeches or verbal explication; and therefore as *veritas non quærit angulos*, so also she needs not the expression of eloquent words and refined sentences or phrases to illustrate it and make it more perspicuous in the eyes of wise and learned men, being that it is not unknown to them, that she does conquer all, for she is the bright splendour or emanation which springs from the omnipotent and eternal fountain; she enlightens all, she acts essentially in all, and over all, and reveals herself in effect to all; and therefore she is so manifest in all her works, that she needs not any golden-tongued Oratour, nor smooth and methodical Rhetorician, or lip-learned Philosopher to do her honour in the expression of her excellency . . . And therefore I will say as the Prophet *David* did, I will sing the truth under the shadow of thy wings. O Lord, in thee do I put my trust; keep and preserve me from mine enemies, for thy mercys sake, *AMEN*.

Notes

1. Elsewhere in the *Mosaicall Philosophy*, Fludd explains the scheme slightly differently:

    So that we see, that as well there are good beams or benign spirits, which by a secret hidden Emanation, so stream forth from the fountain of the winds, namely, those which are poured out by JEHOVAS benign Attributes as are EL by Jupiter, ELOACH by the Sun, SADAI by Venus, and then Michael, Raphæl, Gabriel, etc., do by their Legions execute Gods Will. And there are bad and corrupt or privative emissions of spirits, of a contrary fortitude from the winds, which have their Origin from ELOHIM, ELOHIM GIBBOR, and ADONAI, which make Saturn, Mars and the Moon their store house . . .

    By reason of these spirits of a contrary fortitude in the air, sometimes good and propitious events befall the creatures of this lower world, namely when the good spirits reign, and wholesome winds do blow, which happens when the benign Stars and Planets have dominion in Heaven and consequently their influences below. And again, sometimes bad and disastrous accidents, armed with privative and destructive effects, befall the creatures of the Elementary region, by reason of severe emissions of beams from the winds, which animate those evil spirits, that in infinity multitudes do hover, though invisible, in the air, who are rejoiced and revived by the blasts which issue from the stations of their cruel Princes . . . the whole air is replenished as well with spirits of darkness as with spirits of light. (Book Two, Chapter IV.)

Bibliography

Outline of the *Utriusque cosmi . . . historia*

**Volume I
The History
of the
Macrocosm**

Tractate I (UCH I, 1)
Utriusque cosmi . . . historia.
(Oppenheim: de Bry, 1617)

Tractate II (UCH I, 2)
Tractatus Secondus, De Naturae Simia
(Oppenheim: de Bry, 1618)

**Volume II
The History
of the
Microcosm**

Tractate I

Section I (UCH II, 1, i)
Tomus Secundus De Supernaturali . . .
(Oppenheim: de Bry, 1619)

Section II (UCH II, 1, ii)
Tomi Secundi Tractatus Primi . . .
(Oppenheim: de Bry, 1620?)

Tractate II

Section I

Portions I & II (UCH II, 2, i, 1 & 2)
Tomi Secundi . . . De Praeternaturali . . .
(Frankfurt: de Bry, 1621)

Portion III (UCH II, 2, i, 3)
Anatomiae Amphitheatrum
(Frankfurt: de Bry, 1623)

Portion IV (UCH Ii, 2, i, 4)
Philosophia sacra
(Frankfurt: Officina Bryana, 1626)

Sections II & III (not published)

Tractate III (not published)

Outline of the *Medicina Catholica*

Volume I

Tractate I (MC I, 1)
Medicina Catholica, Sanitatis Mysterium
(Frankfurt: William Fitzer, 1629)

Tractate II

Section I (MC I, 2, i)
Integrum Morborum Mysterium
(Frankfurt: William Fitzer, 1631)

Section II (MC I, 2, ii)
KATHOLIKON MEDICORUM KATOPTRON
(Frankfurt: William Fitzer, 1631)

Section II, Portion III, Part III
Pulsus   (MC I, 2, ii, 3, 3)
(Frankfurt: William Fitzer, 1631?)

Primary Sources

Fludd Manuscripts

Cambridge, Trinity College Library, Western MS, 1150, Fludd, 'A Philosophical Key'. Published edition: Allen G. Debus, *Robert Fludd and his Philosophical Key* (Science History Publications, New York, 1979).

London, British Library, Royal MS, 12 C ii, Fludd, 'Declaratio Brevis'. Published edition: tr. William H. Huffman and Robert A. Seelinger, Jr, introduction by William H. Huffman, 'Robert Fludd's "Declaratio Brevis" to James I' in *Ambix* 25 (1978), 69-92.

New Haven, Yale University Library, Miscellaneous Manuscript 170, Filmer MS. 3, 'Musical compositions by Fludd'. See Todd Barton, 'Robert Fludd's Temple of Music: A Description and Commentary', unpublished MA thesis, University of Oregon, 1978; Appendix III, 202-210.

Oxford, Bodleian Library, MS. Ashmole 766 and MS. Ashmole 1507, Fludd, 'Truth's Golden Harrow'. Published edition: C. H. Josten, 'Truth's Golden Harrow' in *Ambix* 3 (1948), 91-150.

Fludd's Printed Works

Fludd, Robert, *Anatomiae amphitheatrum effigie triplici, more et conditione varia designatum* (Johann-Theodor de Bry, Frankfurt, 1623) (UCH II,2,i,3). This is a continuation of *The History of the Microcosm*; see below and outline of *Utriusque cosmi . . . historia*.

——, *Apologia compendiaria, Fraternitatem de Rosea Cruce suspicionis et infamiae maculis aspersam, veritas quasi Fluctibus abluens et abstergens* (Gottfried Basson, Leiden, 1616). Fludd's first publication, rushed into print as a defence of the Rosicrucians against the attack by Andreas Libavius in his *Analysis Confessionis Fraternitatis de Rosea Cruce* of 1615.

——, *Clavis Philosophiae et Alchemiae Fluddanae* (William Fitzer, Frankfurt, 1633). Fludd's final reply to Mersenne and Gassendi.

——, *Discursus de unguento armario* in Rattray, Sylvestri, ed., *Theatrum Sympatheticum Auctum, exhibens varios authores* (Johan. Andrea Endtera & Wolfgang Junioris Haeredes, Nüremberg, 1622).

——, *Doctor Fludds Answer unto M. Foster, or the squeesing of Parson Fosters Sponge, ordained by him for the wiping away of the weapon-salve* (Nathanael Butter, London, 1631).

——, *Responsum ad Hoplocrisma-spongum M. Fosteri . . .* (Petrus Rammazenius, Gouda, 1638). Latin translation of above.

——, *Medicina Catholica, seu mysticum artis medicandi sacarum. In tomos divisum duos* (William Fitzer, Frankfurt, 1629) (M.C. I,1). This edition contains the Tractatus Primus of the Tomus Primus bearing the title 'Sanitatis Mysterium'. The remainder of the *Medicina Catholica* is under the title *Integrum Morborum Mysterium*.

——, *Integrum Morborum Mysterium: sive Medicinae Catholicae tomi primi tractatus secundus in Sectiones distributus duas* (Wolfgang Hofmann, sold by William Fitzer, Frankfurt, 1631) (M.C. I, 2, i). This continuation of the *Medicina Catholica* contains three treatises belonging to the Tomus Primus and a fold-out schematic of the Tomus Secundus which treatise is the Sectio Prima of the Tractatus Secundus, 'Integrum morborum, seu meteororum insalubrium mysterium'. Sectio Secundus in KATHOLIKON MEDICORUM KATOPTRON, sive tomi primi, tractatus secundi, sectio secunda, 'De Morborum Signis' and bears the date 1631 on the title page (M.C. I, 2, ii). The third treatise with a separate title page has the title 'PULSUS, seu nova et arcana Pulsuum historia, esacro fonte radicaliter extracta, nec non medicorum ethnicorum dictis & authoritate comprobata, hoc est, portionis tertiae pars tertia of the Sectio Secundus de Pulsuum Scientia' (n.p., n.d. [p. 93 bears the date 1629]) (M.C. I, 2, ii, 3, 3). The Sectio Tertius was never printed. The fold-out schematic for the Tomus Secundus is bound at the end with the title 'Medicamentosum Apollinis Oraculum, hoc est, Medicinae Catholicae, sue mysticae medicandi artis, Tomus Secundus' (Typus excudebatur Wolfgangi Hofmanni, n.p., 1630). This entire volume was reissued under the title *Meteorum insalubrium mysterium* (L. Bourgeat, Moguntiae [Mainz], 1682).

——, *Monochordum Mundi Symphoniacum, seu. Replicatio Roberti Fludd … ad Apologiam … Iohannis Kepleri …* (Erasmus Kempfer, Johann-Theodor de Bry, Frankfurt, 1623). Fludd's final reply to Kepler.

——, *Philosophia Moysaica* (Petrus Rammazenius, Gouda, 1638). Fludd's last work, published posthumously; see below.

——, *Mosaicall Philosophy* (Humphrey Moseley, London, 1659). English translation of above. Reprint edition of Books One and Two of the Second Section: Adam McLean, ed., Magnum Opus Hermetic Sourceworks No. 2 (Edinburgh, 1979). Internal evidence on p.232 suggests Fludd wrote most of this work in 1630 or 1631, although on p.287 he mentions visiting a patient in 1637, so the work may have been finished only shortly before his death.

——, *Philosophia sacra et vere Christiana seu Metorologia Cosmica* (Officina Bryana, Frankfurt, 1626) (UCH I, 2, i, 4). This volume is also a continuation of *The History of the Microcosm*; see below and outline of the *Utriusque cosmi … historia*.

——, *Sophie cum moria certamen, in quo, lapis lydius à falso structore, Fr. Marino Mersenno, Monacho, reprobatus, celeberrima Volumnis sui Babylonici (in Genesi) figmenta accurate examinat* (n.p., 1629). (Bound with the *Philosophia Sacra* above, with separate pagination.) Fludd's first reply to Mersenne.

——, *Tractatus apologeticus integritatem Societatis de Rosea Cruce defendens* (Godfrey Basson, Ludguni Batavorum [Leyden], 1617). Fludd's longer defence of the Rosicrucians which followed through on the outline proposed in the *Apologia compendiaria* the year before. There is a German translation of this work by Adam Michael Birkholz [Ada Mah Booz, pseudonym], *Schutzschrift für die Æchtheit der Rosenkreutzergesellschaft* (A. E. Boehme, Leipzig, 1782).

——, *Tractatus theologo-philosophicus, In Libros tres distributus, Quorum I de Vita, II de Morte, III, de Resurrectione . . . à Rudolfo Otreb Britanno* (Johann-Theodor de Bry, Oppenheim, 1617). An early work under an anagram pseudonym.

——, *Utriusque cosmi maioris scilicet et minoris metaphysica, physica atque technica historia . . . Tomus primus, De Macrocosmi Historia in duos tractatus diuisa* (Johann-Theodor de Bry, Oppenheim, 1617) (UCH I, 1). This first issue of *The Technical Metaphysical and Physical History of the Macrocosm and Microcosm* contains the Tractate One of Volume One (The History of the Macrocosm), 'De metaphysico macrocosmi et creaturarum illus ortu et de physico macrocosmi in generatione et corruptione progressu'. English translation of Books One and Two: Fludd, *The Origin and Structure of the Cosmos*, tr. Patricia Tahil, Magnum Opus Hermetic Sourceworks No. 13 (Edinburgh, 1982).

——, *Tractatus Secundus, De Naturae Simia seu Techinca macrocosmi historia, in partes undecim divisa* (Johann-Theodor de Bry, in Nobili Oppenheim, 1618, second edition, 1624) (UCH I, 2). This second tractate of *The History of the Macrocosm* contains eleven essays on man's arts, through the use of which he becomes the 'ape of nature'. They are: I. Universal Arithmetic; II. The Temple of Music; II. Geometry; IV. Optics; V. The Pictorial Arts; VI. The Military Arts; VII. Motion; VII. Time; IX. Cosmography; X. Astrology; XI. Geomancy. Reissued: Johann-Theodor de Bry, Frankfurt, 1624. The section on geomancy appeared under the title 'De animae intellectualis scientia, seu, Geomantia hominibus appropriata' in *Fasciculus Geomanticus, in quo varia variorum opera geomantica continentur* (Verona, 1687; second edition, Verona, 1704). The same section was translated into French by Pierre Vincent Piobb, *Traité de Géomancie (de geomantia)* (Dangles, Paris, 1947). Piobb also translated the section on astrology: *Étude du macrocosme, annotée et traduite pour la première fois par Pierre Piobb. Traité d'astrologie générale (De astrologia)* (H. Daragon, Paris, 1907).

——, *Tomus Secundus de supernaturali, naturali, praeternaturali et contranaturali micrososmi historia, in tractatus tres distributa* (Johann-Theodor de Bry, Oppenheim, 1619) (UCH II, 1, i). This is Section One of Tractate One of Volume II (*The History of the Microcosm*). It is divided into 13 books dealing with divine harmonics as they relate to man, the microcosm. At the end of Book IX there is an errata sheet and index. Immediately following is a title page and new pagination for the second section of the first treatise, bearing the title *Tomi Secundi, tractatus primi, sectio secunda, de technica Microcosmi historia, in Portiones VII divisa* (n.p., n.d.) (UCH II, 2, i). Section II contains essays on Prophesy, Geomancy, The Art of Memory, Astrology, Physiognomy, Chiromancy and the Pyramid. Beginning with p.204 and continuing the new pagination, Books X–XIII of the Sectio Prima appear, ending on p.277.

——, *Tomi secundi tractatus secundus, de praeternaturali utriusque mundi historia. In sectiones tres divisa* . . . (Johann-Theodor de Bry, Francofurti, 1621) (UCH II, 2, i, 1 & 2). This continuation of the Microcosm History contains Portions I and II of Section I; Portions III and IV appeared under separate titles as *Anatomiae Amphitheatrum* (UCH II, 2, i, 3) and Philosophia Sacra (UCH II, 2, i, 4). Sections II and II never appeared.

——, *Veritatis proscenium* . . . *seu demonstratio quaedam analytica, in qua cuilibet comparationis particulae, in appendice quadam J. Kepplero, nuper in fine Harmoniae suae Mundanae edita* . . . (Johann-Theodor de Bry, Frankfurt, 1621). Fludd's first reply to Kepler's objections in the Appendix to his *Harmonices mundi* of 1619.

Selected Contemporary Works

Bacon, Francis, *The Philosophical Works of Francis Bacon*, ed. John M. Robertson (Routledge, London, 1905). It is obvious that Bacon's ideas of a new science not based on religion or the ancients were not compatible with Fludd's. The two were acquainted, but never mentioned one another.

Dee, John, *A True and Faithful relation of what passed for many years between Dr. John Dee and some spirits* (T. Garthwait, London, 1659). Although there are shared interests and outlook between the two, there is no direct link between Dee and Fludd; Fludd would not have anything to do with angel-summoning or 'vulgar mathematics'.

Foster, William, *Hoplocrisma Spongus, or a Sponge to wipe away the weapon-salve* (Thomas Cotes, London, 1631). Parson Foster's attack on Fludd's support of the use of the weapon-salve claiming it worked through diabolical means.

Frizius, Joachim, *Summum Bonum* (n.p., Frankfurt, 1629). Another tract defending Fludd against Mersenne. A number of authors have claimed that Fludd wrote this tract, but he specifically denied having done so. A German translation of Book IV is in F. Freudenberg, *Paracelsus und Fludd* (Hermann Barsdorf Verlag, Berlin, 1918), 233–71; and an English translation of Freudenberg above is found in Paul M. Allen, ed., *A Christian Rosenkreutz Anthology* (Rudolf Steiner, Blauvelt, NY, 1968), 349–79. In the latter Fludd is named as the author, and a number of plates from his books are included in the anthology.

Gassendi, Pierre, *Epistolica exercitatio in qua principia philosophiae Roberti Fluddi, medici, releguntur, et ad recentes illius libros adversus R. P. F. Marinum Mersennum . . . respondetur* (S. Cramoisy, Paris, 1630). Gassendi's examination of Fludd's works, done at Mersenne's request.

Gilbert, William, *De magnete* (P. Short, London, 1600). Fludd used Gilbert's work and that of Ridley (below) to refute Aristotle and prove his theories that there is action between two bodies at a distance.

Harvey, William, *De motu locali animalium*, ed. and trans.

Gwenneth Whitteridge (The University Press, Cambridge, 1959). Fludd was the first author to agree with Harvey's theory of the circulation of the blood in print.

Kepler, Johannes, *Harmonices mundi* (Johannes Planc, Linz, 1619). The Appendix takes issue with Fludd's harmonic scheme of the universe.

——, *Mysterium Cosmographicum* (Erasmus Kempfer, Frankfurt, 1621). In this second edition of this work (the first appeared in 1597), Kepler's reply to Fludd's *Veritatis proscenium* is included in a tract entitled *Johannis Kepleri pro suo Opere Harmonices Mundi Apologia adversus Demonstrationem Analyticam Cl. V. D. Roberti de Fluctibus, Medici Oxoniensis.*

Lanovius (François de la Noue), *Ad Reverendum Patrem Marinum Mersennem Francisci Lanovii Judicium de Roberto Fluddo* (Paris, 1630). Lanovius takes Mersenne's side in the controversy with Fludd.

Mersenne, Marin, *Quaestiones celeberrimae in Genesim* (Paris, 1623). Includes Mersenne's attack on Fludd.

Ridley, Mark, *A Short Treatise of Magneticall Bodies and Motions* (Nicholas Okes, London, 1613). See note on Gilbert above.

Scot, Patrick, *The Tillage of Light* (William Lee, London, 1623). Scot maintained that the Philosopher's Stone was not a material substance. Fludd replied with the unpublished manuscript 'Truth's Golden Harrow'.

Selden, John, *Titles of Honour* (W. Stansby for I. Helme, London, 1614). The noted legal scholar and antiquarian praised Fludd in the dedication (to a friend) for curing his sickness.

Thornborough, John, *ΑΙΘΟΘΕΩΡΙΚΟΣ*, sive *Nihil, aliquid, omnia, antiquorum sapientium vivis colorbus depicta* ... (I. Lichfield and I. Short, Oxford, 1621). This tract by the Bishop of Worcester, a friend of Fludd's, contains marginal references to Fludd's *Tractatus theologo-philosophicus* of 1617.

Secondary Sources

Books on Fludd

Craven, James Brown, *Doctor Robert Fludd (Robertus de Fluctibus), The English Rosicrucian: Life and Writings* (William Peace & Son, Kirkwall, 1902). Contains a now outdated biography but a good summary of Fludd's works and a detailed bibliography.

Godwin, Joscelyn, *Robert Fludd: Hermetic Philosopher and Surveyor of Two Worlds* (Shambhala, Boulder, Colorado, and Thames and Hudson, London, 1979). Excellent collection of Fludd's illustrations with detailed explanations and an important bibliography.

Huffman, William H., *Robert Fludd and the End of the Renaissance* (Routledge, London and New York, 1988). The first monograph on Fludd since the revival of interest in his work by Dame Frances Yates, Allen G. Debus and others. Contains information on his life, works, the Rosicrucians, his will and a detailed bibliography.

Hutin, Serge, *Robert Fludd (1574-1637): Alchimiste et Philosophe Rosicrucien* (Omnium Litteraire, Paris, 1971). This is basically Hutin's 1951 thesis for an *Elève diplomé* and does not therefore take into account any modern scholarship on Fludd.

Selected Other Books and Articles

Ammann, Peter J., 'The Musical Theory and Philosophy of Robert Fludd' in *Journal of the Warburg and Courtauld Institutes* 30 (1967), 198-227.

Barton, Todd, 'Robert Fludd's *Temple of Music*: A Description and Commentary', unpublished MA thesis, University of Oregon, 1978.

Bonelli, M. L. R. and Shea, W. R., eds, *Reason, Experiment and Mysticism in the Scientific Revolution* (Science History Publications, New York, 1975).

Copenhaver, Brian P., 'Natural Magic, Hermeticism, and

Occultism in Early Modern Science' in David C. Lindberg and Robert S. Westman, *Reappraisals of the Scientific Revolution* (Cambridge University Press, Cambridge, 1990).

Debus, Allen G., 'The Chemical Debates of the Seventeenth Century' in *Reason, Experiment and Mysticism in the Scientific Revolution* (Science History Publications, New York, 1975).

——, *The Chemical Dream of the Renaissance* (W. Heffer and Sons, Cambridge, 1968).

——, *The Chemical Philosophy: Paracelsian Science and Medicine in the Sixteenth and Seventeenth Centuries* (Science History Publications, New York, 1977).

——, *The English Paracelsians* (Franklin Watts, New York, 1965).

——, *Man and Nature in the Renaissance* (Cambridge University Press, Cambridge, 1978).

——, *Science and Education in the Seventeenth Century: The Webster-Ward Debate* (Science History Publications, New York, 1970).

Allen Debus has also written numerous articles relating to Fludd. It is due to his work from the perspective of the history of science and medicine that modern scholarship established Fludd's importance in the early seventeenth century.

Dictionary of National Biography, s.v. 'Fludd, Robert' by Alexander Gordon.

Dictionary of Scientific Biography, s.v. 'Fludd, Robert' by Allen G. Debus.

Encyclopedia of Philosophy, s.v. 'Fludd, Robert' by John Passmore.

Godwin, Joscelyn, 'Instruments in Robert Fludd's *Utriusque Cosmi . . . Historia*' in *Galpin Society Journal* 26 (1973), 2-14.

——, ed., *Music, Mysticism and Magic: A Sourcebook* (Penguin, Harmondsworth, 1986).

——, 'Robert Fludd on the Lute and Pandora' in *Lute Society Journal* 15 (1973), 11-19.

High Matter, Dark Language: The Philosophy of Robert Fludd (1574-1637). An Exhibition at the Wellcome Institute for the History of Medicine. Catalogue (The Wellcome Institute for the

History of Medicine, London, 1984).

Josten, C. H., 'Robert Fludd's "Philosophical Key" and his Alchemical Experiment on Wheat' in *Ambix* 11 (1963), 1–26.

——, 'Robert Fludd's Theory of Geomancy and his Experiences at Avignon in the Winter of 1601 to 1602' in *Journal of the Warburg and Courtauld Institutes* 27 (1964), 327–35.

——, 'Truth's Golden Harrow: An Unpublished Alchemical Treatise of Robert Fludd' in *Ambix* 3 (1949), 91–150.

The above are three important articles about Fludd and his philosophy.

Lindberg, David C. and Robert S. Westman, *Reappraisals of the Scientific Revolution* (Cambridge University Press, Cambridge, 1990).

McLean, Adam, *The Western Mandala* (Hermetic Research Series, Edinburgh, 1983, distributed by Phanes Press). Adam McLean has done much over the last few years to republish Hermetic Renaissance texts and write articles and books about them.

Merkel, Ingrid, and Allen G. Debus, eds., *Hermeticism and the Renaissance: Intellectual History and the Occult in Early Modern Europe* (Folger Shakespeare Library, Washington, D.C., Associated University Presses, London and Toronto, 1988).

Pagel, Walter, *Paracelsus: An Introduction to Philosophical Medicine in the Era of the Renaissance* (S. Karger, Basel and New York, 1958).

——, 'Religious Motives in the Medical Biology of the Seventeenth Century' in *Bulletin of the Intitute of the History of the Medicine* 3 (1935), 97–312.

Pauli, Wolfgang, 'The Influence of Archetypal Ideas on the Scientific Theories of Kepler' in *The Interpretation of Nature and the Psyche*, Bollingen Series 51 (Pantheon Books, New York, 1955).

Singleton, C. S., ed., *Art, Science and History in the Renaissance* (Johns Hopkins Press, Baltimore and London, 1967).

Thorndike, Lynn, *A History of Magic and Experimental Science*, 8 vols (Columbia University Press, New York, 1923–58).

Vickers, Brian, ed., *Occult and Scientific Mentalities in the Renaissance* (Cambridge University Press, Cambridge, 1984).

Waite, Arthur Edward, *The Real History of the Rosicrucians* (Redway, London, 1887).

——, *The Brotherhood of the Rosy Cross* (1924; reprint University Books, New Hyde Park, NY, 1961).

——, 'Robert Fludd: Philosopher and Occultist' in *The Occult Review* 15 (1912), 79–84.

——, 'Haunts of the English Mystics' in *Square and Compass* 47 (1933), 11–15.

Walker, D. P., *Spiritual and Demonic Magic from Ficino to Campanella* (The Warburg Institute, London, 1958; reprints: Kraus, 1969; University of Notre Dame Press, 1975).

——, *The Ancient Theology* (Cornell University Press, Ithaca, 1972).

——, 'The Astral Body in Renaissance Medicine' in *Journal of the Warburg and Courtauld Institutes* 21 (1958), 119–33.

Webster, Charles, *From Paracelsus to Newton: Magic and the Making of Modern Science* (Cambridge University Press, Cambridge, 1982).

Westman, Robert S., 'Magical Reform and Astronomical Reform: The Yates Thesis Reconsidered' in *Hermeticism and the Scientific Revolution* (William Andrews Clark Memorial Library, Los Angeles, 1977).

——, 'Nature, art and psyche: Jung, Pauli, and the Kepler-Fludd polemic' in *Occult and Scientific Mentalities in the Renaissance*, ed. Brian Vickers (Cambridge University Press, Cambridge, 1984). This is an excellent paper that summarizes the debate, adds important new information and insight, and should be read in conjunction with Wolfgang Pauli's paper excerpted in this volume.

Yates, Frances A., *Giordano Bruno and the Hermetic Tradition* (Routledge and Kegan Paul, London; University of Chicago Press, Chicago, 1964).

——, *The Art of Memory* (Routledge and Kegan Paul, London; University of Chicago Press, Chicago, 1966).

——, *Theatre of the World* (Routledge and Kegan Paul, London;

University of Chicago Press, Chicago, 1969).

——, *The Rosicrucian Enlightenment* (Routledge and Kegan Paul, London and Boston, 1972).

——, *Shakespeare's Last Plays* (Routledge and Kegan Paul, London, 1975).

——, *The Occult Philosophy in the Elizabethan Age* (Routledge and Kegan Paul, London and Boston, 1979).

——, 'The Hermetic Tradition in Renaissance Science' in *Art, Science and History in the Renaissance*, ed. Charles Singleton (Johns Hopkins Press, Baltimore and London, 1967).

——, 'The Stage in Robert Fludd's Memory System' in *Shakespeare Studies* 3 (1967), 138-66.

Dame Frances Yates' intriguing writings on the history of Hermetic and magical thought did much to rekindle an interest in this era from a new perspective. Her speculations have sparked much debate, particularly about the 'Yates thesis': that Early Modern Hermetic thinkers were a key bridge from ancient to modern science. She was the first modern scholar to point out the importance of the *Corpus Hermeticum* for those in the Neoplatonic camp, starting with Ficino in Florence; this was most particularly true for Robert Fludd, as she amply shows. She also dealt with other aspects of Fludd, e.g. his memory system as used in the theatre, and makes use of many of his illustrations. Some of her speculations still invite research, such as the idea that the Rosicrucians were a secret movement to support the Protestant League on the Continent before the Thirty Years War.

Index